imagine 02 – DEFLATEABLES

Delft University of Technology, Faculty of Architecture,
Chair of Design of Constructions

imagine 02

SERIES EDITED BY
Ulrich Knaack
Tillmann Klein
Marcel Bilow

DEFLATE-ABLES

Ulrich Knaack
Tillmann Klein
Marcel Bilow

With contributions by:
Arie Bergsma
Andrew Borgart
Raymond van Sabben
Peter van Swieten

010 Publishers Rotterdam 2008

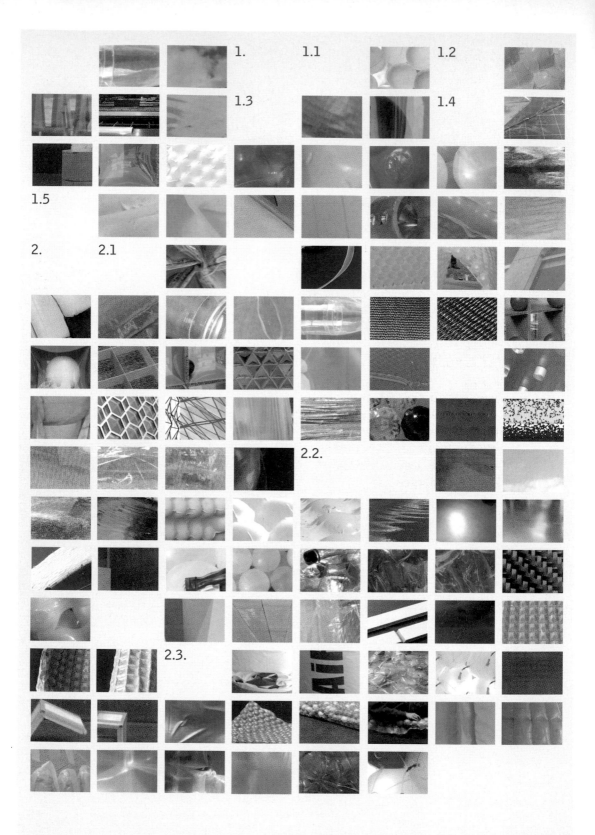

1.
1.1
1.2
1.3
1.4
1.5
2.
2.1
2.2.
2.3.

CONTENTS

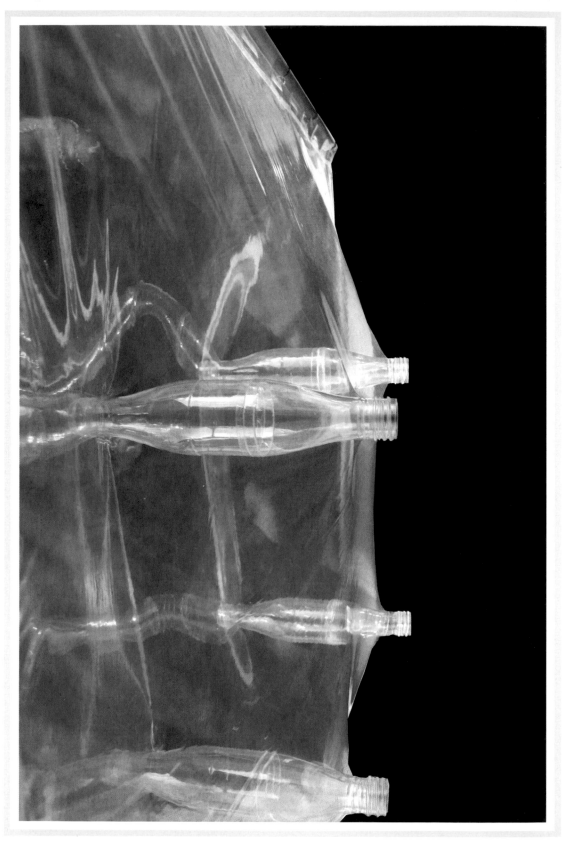

INTRODUCTION –
ENHANCING THE BOUNDARIES

THE IDEA

Deflated constructions are part of the family of pneumatic constructions. The combination of air-pressure and textiles has been used in the field of building construction for quite some time. Famous examples such as the Eden Project in Cornwall and the new Munich Football stadium have made this building type very fashionable. Current pneumatic constructions present an image of elegance and performance due to their lightweight capabilities and potential for transparency.

The field of deflateables is still largely undiscovered, certainly when compared to that of inflated constructions. Deflateables follow the inversed principle. After minor attempts in the 1970s, the topic of deflated constructions has recently become a major focus of research in building technology. One of the reasons for this lies in the increasing possibilities of membrane technology. Our Chair of Design of Constructions has studied the topic and developed an overview of current developments. What are deflated constructions precisely, and what are their structural and physical properties? We explored potential areas of application in a large number of projects. The projects that seemed promising were researched on a deeper level. We integrated the topic into our research and education program, and it directly aroused the interest of architects. The potential of this technology for architectural expression has not yet been fully exploited.

With 'Deflateables', the second in the series of IMAGINE books, we hope to able to share our enthusiasm for this technology with our readers. We wish to present a snapshot of our work in this particular field, share our knowledge, and encourage others to participate in our explorations.

WHO

Within the Technical University of Delft / Faculty of Architecture, the Façade Research Group was set up as an impulse program to advance academic research in a field in which European engineering already is a world leader. The group is anchored within the Faculty, in the Building Technology / Chair of Design of Constructions headed by Prof. Dr Ulrich Knaack. The starting point for the formation of the group was an analysis of existing strategic, technical and design-related knowledge, and of the different mechanisms of the market and established research. Then these results had to be linked. At present, several research projects and PhD theses are being studied – supported by universities and industrial partners. The PhD students Lidia Badarnah, Marcel Bilow, Thiemo Ebbert, Daan Rietbergen, and the researchers Ari Bergsma and Tillmann Klein, who also leads the research group, are currently involved in the program.

The group is linked to the "Material und Gebäudehülle" (Material and
Building Cladding) research theme of the Detmolder Schule für Architektur
und Innenarchitektur in Germany, where Ulrich Knaack holds the chair of
"Entwerfen und Konstruieren" (Design and Construction), and also partici-
pates in the "Forschungsschwerpunkt Material und Gebäudehülle"
(Material and Building Cladding specialist research) research group along
with Marcel Bilow.

WHAT

Façade technology of the 20th century is related to the dissolution of the
massive wall into a separation of structure and façade. Looking at the
development of façade technology nowadays, after 60 years of curtain wall
systems, 30 years of element-façade systems, and 10 years of experience
with the integration of environmental services and double-skin façades, we
must conclude that the peak of optimization has been reached. No further
technological advance can be expected by continuing the policy of adding
extra layers for each additional technical function.
In combining different disciplines and technologies, we pursue the evolution
of façades – or "skins", to apply a more relevant word – by using alternative,
new and perhaps even embryonic technologies that have not been fully
developed yet, but merely mentioned somewhere. The results are organized
according to topic, and present the main idea by means of sketches,
pictures and explanatory text. Keywords were used to organize the ideas in
a database, and are highlighted in this book for quick orientation.

HOW

The prevailing approach to this research was to try to work with inspiration
and an open mind. To achieve this, we needed not only an interesting and
divers group of people who displayed mutual trust, but we also had to use
various methods to create a supportive environment.
All materials have been developed as part of the "Deflateables" program or
in courses and collaborations linked to the program. We would like to thank
all of our colleagues and students for providing us with their materials;
especially Raymond van Sabben, who worked on this project as a student
assistant, Linda Hildebrand for her work in collecting the material for
this publication and all the other people who supported projects that, due
to the shear volume, could not be included in this volume.

Prof. Dr.-Ing. Ulrich Knaack

1. BASICS

1.1. PROGRAM SETUP

The entire Deflateables project started by coincidence. We were able to purchase a used vacuum pump from a friendly supplier. The new tool inspired the whole crew to take a deeper look at the potential of this technology. What are the characteristics of vacuum and how can it be used in building construction?

Deflateables belong to the family of pneumatic constructions. Their specialty is the structural use of low-pressure. A survey about existing work in this field showed that very little research has been done, and that deflated structures are far less applied in real systems than their overpressure counterparts. The 1970s witnessed the first steps to explore this type of construction. The first work we found was a research project called "Vacuumatics", carried out at the University of Belfast [Gilbert, 1970]. The work of the Institut für Leichte Flächentragwerke (IL) in Stuttgart was very inspiring. The institute did some fundamental research in the field of pneumatic constructions. The book entitled *IL 9* [Burkhardt, 1977] explains some general examples of vacuum-stiffened structures. In the book entitled *Pneumatic Structures* [Herzog, 1977], a survey of practical and theoretical aspects of pneumatic constructions is given, representing the state-of-the-art of this time. Currently, the University of Stuttgart is researching practical applications [Sobek, 2007] and the University of Technology Eindhoven is executing a research project in this field [Huijben, 2007].

The goal of this book is to give a general overview of this topic and to show a great variety of possibilities. It also contains a collection of the work that has been performed by us so far. Some contributions are presented as basic ideas, "imaginations" of possible applications; some are worked out to a detailed level with all necessary calculations.

The content is divided into the following chapters:
- A short history of deflateables
- General, theoretic introductions to mechanical and building physical behavior and practical advice
- A collection of imaginations with a subdivision into the topics of façades, structures, furniture and interiors
- The research was closely linked to the educational program of the Faculty and one chapter describes the experiences of the first semester of the Master's degree program of Building Technology – designing prototypes.

With this book we would like to motivate others to pick up on these ideas and continue the research.

1 The principle of Torricelli's mercury barometer
2 Otto von Guericke's Magdeburger Halbkugel

1.2. SHORT HISTORY OF DEFLATEABLES

BUILDING SOMETHING WITH THE ABSENCE OF MATTER

Simply speaking, vacuum can be defined as a space with no matter inside of it, although we know today that this is physically impossible.

The Romans wee the first to name the phenomenon: "Vacuum is where there is nothing", using the Latin word "vacuus", meaning empty. From the earliest times until today, vacuum has been the focus of many philosophical discussions. Aristotle doubted the existence of vacuum – nothing cannot be something – and in medieval times the absence of all things was regarded as impossible, since that would mean the absence of God.

The Galilean student Evangelista Torricelli (1608-1647) was the first to demonstrate vacuum, created by experiment. He filled a one-sided, calibrated glass tube with mercury. When he put the glass with the open end pointing downward in a bowl filled with the same material, the level of the silver liquid dropped down, leaving a vacuum at the top end of the tube. (fig. 1) The height of the mercury column changes with the surrounding atmospheric pressure. The Torricelli vacuum is named after him, as well as the unit 'Torr', a unit of pressure. He invented the first barometer. For more information on `Torr`, see thee chapter on "Building physical properties".

Otto von Guericke (1602-1686) also studied the vacuum. He invented the first vacuum pump. His most successful experiment was joining the "Magdeburger Halbkugel" in 1663. He joined two copper hemispheres together by pumping the air out of the space between. Eight horses were harnessed to each hemisphere and tried to pull the hemispheres apart, unsuccessfully. (fig. 2) The two hemispheres were only fixed by the absence of air, or by the pressure difference between the inside and the outside.

After centuries of philosophical and physical discussions, it is easy for us to accept vacuum as something real. Vacuum is part of our everyday lives. The light bulb is just one example of many: Edison used a vacuum in the light bulb to prevent the filament from being burned by oxygen.

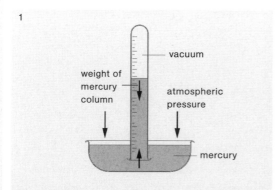

1

vacuum

weight of mercury column

atmospheric pressure

mercury

2

3 Vacuum lifting device for glass panes
4 Manufacturing panels for the automotive industry
5 Scheme for vacuum injection molding

Other well known examples are vacuum-packaging to preserve food for extended periods of time. Coffee beans are compressed and interlock. The inherent friction provides certain stability. A coffee pack becomes rigid and maintains its shape when the air is drawn out. Another example of the use of vacuum is the thermos flask. The insulation value is higher than that of regular flasks. The absence of air prevents heat transfer by convection. The vacuum cleaner also uses the principle, as the term already implies: by creating a relative low-pressure inside the device, it picks up particles and moves them to the intake opening.

VACUUM IN BUILDING TECHNOLOGY

A vacuum exhibits four basic properties that we can take advantage of in building technology:
- The absence of air inhibits convection.
- The absence of air reduces the transfer of sound waves.
- Vacuum can compress and stabilize incoherent material if wrapped in an air-tight enclosure.
- Vacuum between two materials can cause them to attach due to the pressure difference.

These characteristics are a basic description of the effects vacuum can create; of course they cannot be applied to every situation. However, they are a good starting point to understand the potential of vacuum. Here are some practical examples:

The ability to attach:

The industry has discovered the various advantages of vacuum for transport purposes. For example, huge vacuum pads are used to lift and move heavy cement bags without the risk of damage. And the glass industry uses vacuum technology to move glass panes during production and assembly (fig. 3). Since the lifting device does not hold the panes at the edges, glass panels can then be easily set in the framing structure.

The ability to compress:

Besides being used to hold heavy and hard-to-handle materials, vacuum is also efficiently applied to connect large devices to each other. Units, such as sandwich panels that need to be joined are fixated with hardening foam and covered with an airtight envelope (fig. 4). To create the necessary compressive force for the gluing process, the unit is deflated. Due to the ambient pressure of approximately 1bar, the vacuum can theoretically develop a thrust force of $10t/m^2$.

Another advantage is that air pressure is all-round and therefore the pressure is evenly distributed over the entire surface. Vacuum's flexible usage and the low cost

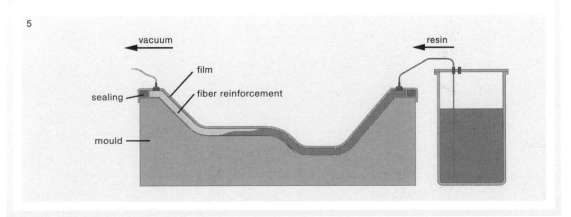

5

vacuum   resin

film

sealing   fiber reinforcement

mould

of material and supplies often make vacuum pressing the most attractive production method.

This method was originally developed in the boat building industry. An additional benefit applied here is that the vacuum causes the resin to penetrate deep into structure (fig. 5). Dry, prefabricated glass-fibre mats are inserted into the original negative mold of the boat hull, until the necessary number of layers are in place. Then the entire form is enveloped with an airtight foil. Pressure tubes are attached to draw out the air. The last step is to apply low-pressure, which sucks the resin into the material, providing a continuous soaking of the fibers and creating a homogenous GFK structure.

The ability to insulate:

The vacuum-insulated panel (VIP) (fig. 6) exploits the above mentioned principle of the thermos flask – the reduction of convection. The production of vacuum insulation panels is simple and similar to the vacuum packing method in the food industry. A VIP essentially consists of a microporous nuclear material that is shrink-wrapped into a highly gas-tight cladding foil in a vacuum chamber. Optimum results rely on low internal convection and thermal conduction of the construction (see the chapter on Building Physical Properties), resulting in 5-10 times thinner panels than standard non-vacuum panels. VIPs are used in the façade industry amongst others. The life span depends on the quality and tightness of the foil. It is important that the panels do not become damaged during the entire lifecycle.

These examples show a range of vacuum applications in the building and other industries. Whereas overpressure has been widely used for structural purposes, the field of vacuum supported structures remains largely undiscovered.

The combination of the above mentioned properties and characteristics provide an interesting field for further research – the goal of this book.

6 Detail of a vacuum-insulated panel
7 Coffee pack

1.3. BUILDING PHYSICAL PROPERTIES

INTRODUCTION

In addition to their potential from a structural point of view, as discussed in other chapters of this book, vacuum structures or vacuum building components can also be used to improve building physical properties such as thermal and acoustic insulation of façade structures. The main purpose of this chapter is to present a global overview of this potential.

This contribution begins with an introduction to the phenomenon of vacuum and its physical characteristics, detailed in a few simple vacuum structures. To link theoretical properties with building practice, the potential of the use of vacuum technology in façade structures is then briefly discussed.

Today, vacuum technology is widely used in several other industries and research areas, such as the space industry, medical industry, electronic industry (older types of television and computer screens, CD-ROM production), nuclear physics, etc. In the building industry, the idea of applying vacuum as a way to improve the building-physical properties of building components and materials is not entirely new either. The previous chapter described some of the existing applications and current research projects.

WHAT IS A VACUUM?

A space containing no matter, a space that can be described as 'completely' empty – containing not even the smallest particles such as atoms, electrons etc. – is called a vacuum. However, an absolute vacuum with these properties has not been observed yet, and probably never will; according to the quantum theory, spaces of volume that are completely empty cannot exist due to the presence of virtual parts. Thus, the concept of an absolute vacuum can be more accurately described as a philosophical definition.

Contrary to physicists – who often use the above definition of vacuum solely for situations with absolute vacuum – technicians generally use the term vacuum for spaces of volume exhibiting a lower pressure than the atmospheric pressure (1 bar). Depending on the vacuum level, it can be divided into four categories:

- Low or coarse vacuum:
  1 bar (= atmospheric pressure) to 10^{-3} bar
- Medium or fine vacuum:
  10^{-3} bar to 10^{-6} bar
- High vacuum: 10^{-6} bar to 10^{-10} bar
- Ultra-high vacuum: less than 10^{-10} bar

Atmospheric pressure = 1 bar = approx. 1013 mbar (approx. $10t/m^2$) = 0.1 MPa.

In the field of vacuum physics, the unit torr is often used (1 Torr = approx. 133 Pa = 1.33×10^{-3} bar).

Ultra-high vacuum (gas pressure less than 10^{-10} bar), for instance, exists in the interstellar space outside our planet. An example of low vacuum (1 bar to 10^{-3} bar) is a thermos flask with a double wall and an evacuated cavity to improve insulation.

Most of the building or façade-related vacuum structures discussed in this book, such as the vacuum bridge or the vacuum cardboard walls, have a vacuum level of approximately 0.2 to 0.3 bar. According to the four categories above, these structures can be defined as coarse vacuum structures. In practice, an ultra-high vacuum can be achieved by means of vacuum pumps; however, it is very difficult to realize pressures lower than 10^{-14} bar. It appears that at these types of pressures, most materials are volatile and easily evaporate or sublimate, thus influencing the state of vacuum.

PHYSICAL PROPERTIES OF A VACUUM
Acoustic behavior

Sound can be defined as a wave motion or vibration that is transmitted through a medium by collisions between adjacent molecules. Sound vibrations need a medium that transmits them. This medium can be gaseous (air), liquid (water) or massive, such as concrete. By successive collisions between molecules, the motion and (kinetic) energy are transmitted. This results in a wave-like transfer of vibrations outward from the source to a receiver, with a gradual decrease in amplitude and thus energy, with increasing distance form the source.

The fact that sound vibrations need a medium to be transmitted also means that if there is no medium – as in a vacuum – there will also be no sound transmission. A classic experiment to demonstrate the effect of air pressure and the degree of vacuum on sound transmission is the experiment with a loudly ticking, mechanical clock with an external clapper underneath a glazed dome. When the dome is filled with air, one will hear and see the clock moving and ticking. However when a vacuum is created in the glazed dome, the ticking can no longer be heard, one will only see the clock moving.

The effect of a vacuum and the relationship between sound transmission – or amount of sound insulation – and air pressure – or air density/degree of vacuum – can be roughly calculated, and will be illustrated by some indicative calculations that were made for a multi-layered façade structure consisting of two transparent EFTE sheets, each with a thickness of 2 mm and a cavity in between (see figure 1). These indicative calculations were carried out with the 'meerlagen model' (multilayer) software of Delft University of Technology (written and programmed by L. Nijs[1]). Please note that these calculations were made for an 'ideal' structure, thus without considering any acoustic leakage such as rigid side connections, support pillars in the cavity, etc. and therefore mainly have a theoretical value by showing physical interdependencies.

Results of the calculations are presented in figure 2, showing the relation between the sound insulation for traffic noise (the Dutch $R_{A,traffic}$) – along the vertical axis – and the depth of a cavity – along the horizontal axis – for several air pressures and degrees of vacuum. Analyzing this figure, it becomes apparent that the sound insulation of the examined structure increases with decreasing air pressure – or increasing vacuum – inside

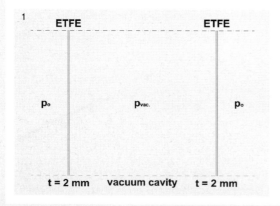

1

ETFE   ETFE

p_0 $p_{vac.}$ p_0

t = 2 mm   vacuum cavity   t = 2 mm

1 Section of vacuum wall
2 Detail of mock-up cardboard façade

See page 24
3 Influence of air pressure on sound insulation
4 Schematic detail of thermal and acoustic separation of deflated façade

the cavity. For a cavity with a coarse vacuum of 0.1 bar, an improvement of up to approximately 10 dB(A) can be realized when compared to an ordinary situation under normal atmospheric pressure. Theoretically, even better sound insulation can be achieved with higher levels of vacuum.

As mentioned above, acoustical leakages were not taken into account in the calculations. Of course, in practice, one cannot ignore these effects, and will have to try to minimize these kinds of leakages. Thus, the actual sound insulation will be significantly less than the theoretical optimum, depending on the detail and acoustical quality of these acoustical bridges.

In the case of the deflated cardboard structure (see figure 2), we could imagine a solution in which a vibration-proof and thermal-insulating material – such as a rubber/cork granulate composite, for example – is glued on cardboard in the center of the cardboard structure (see figure 4).

Thermal behavior

Depending on the type of medium or material, heat transfer of thermal energy in a body can occur through several mechanisms: through thermal convection, radiation, conduction or any combination of these. In a solid material, such as concrete, only conduction can occur. In another example of an air cavity in a masonry cavity wall, all mechanisms of heat transfer can take place. According to the second law of thermodynamics, heat transfer – convection, radiation, conduction or combination – always takes place from an area of higher temperature to an area of lower temperature.

Convection
Convection is heat transfer by the flow of a liquid or gaseous medium, in which the matter is used as the medium for heat transport. For instance, in the atmosphere of the earth, the lower-density warm air rises; whereas the denser cold air sinks, thereby causing convection flow. Heat convection in an air cavity is expressed by formula 1 in figure 5.

Because convection is based on a flow of a liquid or gaseous medium, there will be no heat convection if there is no medium present, as occurs in an absolute vacuum, or if the medium cannot flow. But in practice, the positive effect of vacuum on convection will depend on the degree of vacuum, and thus the amount of medium still available for convection. Due to the fact that completely empty spaces theoretically do not exist, and high vacuums are very difficult to create (especially in building structures), perfect situations with 'zero' convection are

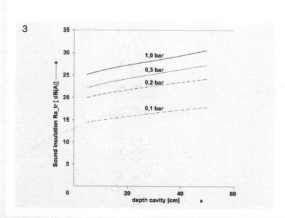

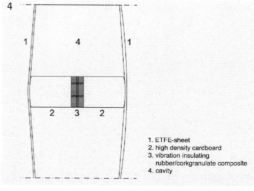

1. ETFE-sheet
2. high density cardboard
3. vibration insulating rubber/corkgranulate composite
4. cavity

not possible. However, depending on the degree of vacuum achieved, heat convection can be significantly reduced.

Conduction

Thermal conduction is the transfer of thermal energy through direct, successive collisions between molecules within a medium or between mediums in direct physical contact. So, contrary to heat convection, heat transfer by conduction occurs without the medium flowing. The conduction properties depend on the type of material and its phase: solid, liquid, or gaseous. Typically, gases are not good conductors; due to the large distance between molecules in a gas, there are fewer collisions, meaning less conduction. Solid metals (e.g. copper) with a more compact stacking of atoms are usually the best conductors of thermal energy.

Besides the properties of the basic material and its phase, the thermal conduction of a building element is directly related to the total thickness of this part. In case of an air cavity: the wider the cavity, the smaller the share of thermal conduction compared to the total heat transfer. The effect of thickness on heat transfer by heat conduction is shown in formula 2 in figure 5. We can conclude that, in the case of vacuum cavities, the relative effect of vacuum on heat con-duction and on heat convection depends on the thickness/depth of the cavity.

Radiation

Radiation is the transfer of heat through electromagnetic radiation, and therefore needs no medium to occur. For instance, solar energy travels from the sun through the vacuum of interstellar space by means of electromagnetic radiation before hitting and warming the earth. On the other hand, the only way that energy can leave earth and travel into space is by radiation, because there is almost no medium for conduction and convection in space. Thermal radiation between two surfaces is expressed by formula 3 in figure 5 (law of Stefan-Boltzmann). In this formula, the emissivity of a surface or coating is expressed by the emissivity coefficient ε. This coefficient is material and frequency dependent. The lower the emissivity value, the lower the radiation. Thus, in order to minimize heat radiation through a cavity, reflective materials or coatings with low values are usually applied to one or both sides of the cavity. A good example is high insulating, double glazing with a metal vacuum-deposited coating (low E–coating) on one of the glass panes enclosing the cavity.

5

Basic formulas heat transfer through cavity:

Convection:

$$q_{cv} = a(d)_c \cdot T_1 \text{-} T_2) = \alpha_c . \Delta T \qquad \text{- 1 -}$$

Conduction:

$$q_{cv} = a(d)_c \cdot T_1 \text{-} T_2) = \alpha_c . \Delta T \qquad \text{- 2 -}$$

Radiation (law of Stefan-Boltzmann):

$$q_{cv} = a(d)_c \cdot T_1 \text{-} T_2) = \alpha_c . \Delta T \qquad \text{- 3 -}$$

Total:

$$q_{cv} = a(d)_c \cdot T_1 \text{-} T_2) = \alpha_c . \Delta T \qquad \text{- 4 -}$$

Explanation symbols:
T_1 = temperature side 1 [K]
T_2 = temperature side 2 [K]
ε_1 = emmisivity coefficient material sid
ε_2 = emmisivity coefficient material sid
α_{cv} = convection coefficient [W/m^{2}K]
α_{cd} = conduction coefficient [W/m^{2}K]
α_r = conduction coefficient [W/m^{2}K]
q_{cv} = heat flow by convection [W/m^2]
q_{cd} = heat flow by conduction [W/m^2]
q_r = heat flow by radiation [W/m^2]

Combination

As already mentioned, in an air cavity all 3 heat transfer processes take place: convection, conduction, and radiation (see figure 6). The total heat transfer through an air cavity – between surfaces with emissivity $\varepsilon = 0.95$ – can be written as formula 4 in figure 5, as a sum of the separate components. For an air cavity, the relation between heat transfer (illustrated by the total heat coefficient α, which is the total heat transfer through a structure per degree K per m^2) and the thickness of the cavity is illustrated in figure 7.

From the physical mechanisms described previously we can conclude that vacuum in a cavity can only exhibit a positive effect on the total heat transfer by means of lower heat conduction and convection. Heat radiation in a cavity cannot be prevented by vacuum. Taking into consideration that the contributions of convection and conduction in the total heat transfer are related to thickness of the cavity, the total effect of vacuum on thermal insulation is relative and also dependent on the thickness of the cavity.

As illustrated in figure 7, the total heat transfer in air cavities with small depths (< 10 mm) is mainly determined by heat conduction. Therefore, with double glazing or fiber-foam insulation panels with small cavities, for example, large improvements

of thermal insulating performance can be achieved by means of vacuum. However, this can only be done if the air pressure in the cavity is low enough. According to Collins and Simko[2] who both conducted in-depth research in vacuum glazing in relation to thermal insulation, it appears that the thermal conductivity of gas is virtually independent of pressure over many orders of magnitude. In their research on vacuum glazing with small cavities of approximately 0.2 mm, significant effects were only achieved for a state of high vacuum (gas pressures lower than 10^{-4} bar). This means that the requirements concerning the degree of vacuum – and its structure – are high and have to be met, before any advantage can be achieved from a conduction point of view.

Furthermore, in the case of large cavity structures, such as the previously discussed ETFE structure for example, the effect of vacuum on the total heat transfer of the structure will be relatively small, as long as the heat radiation is still the major component in the total heat transfer. As can be seen in figure 7, the relative effect of vacuum on heat conduction is almost zero, and the relative effect of vacuum on heat convection is small. As explained previously, for these types of structures, a major improvement in heat insulation can only be achieved by reducing heat

6

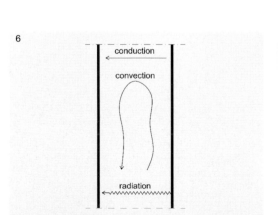

7

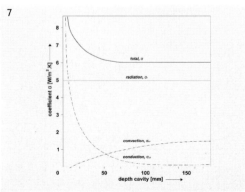

radiation by means of applying low emittance coatings on (one of) the ETFE sheets enclosing the cavity.

When the radiation component of the equation is minimized, a vacuum can have a positive effect on the thermal insulation properties. As discussed before, in case of deep air cavities, a reduction of heat conduction can only be achieved with high or ultra-high vacuum (gas pressures lower than 10^{-4} bar). Due to structural restrictions, these states of high vacuum are not very realistic for building structures and are also difficult to stabilize (leakages, volatility of materials). A more positive effect can be expected by reducing the contribution of heat convection.

WHY VACUUM TECHNOLOGY IN BUILDING INDUSTRY?

In addition to its highly important architectural function, the façade skin of a building may have to fulfill several technical functions: providing air and watertightness, insulating against heat and/or cold, regulating access of daylight and/or solar energy, protecting against outside noise, providing possible provisions for natural ventilation, etc.

The tendency of (European) legislation, driven by the debate about global warming, sustainability, etc., is that requirements on building physical properties of a building, such as insulation and air tightness (from an energy saving point of view) as well as the acoustical performance of building skins (from a user comfort point of view) become increasingly more stringent. In the Netherlands, for instance, the insulating properties of closed façade parts, as well as façade openings, often arise from requirements of the EPC value (Dutch index for energy performance of building, as subscribed in the Dutch Building Decree). The prognosis is that this EPC value will become increasingly stringent in the future. If we consider some of the physical properties of vacuums as described in the previous paragraphs, vacuum technology – if correctly used in properly designed structures – has the potential to improve the insulation as well as the acoustic properties of building structures. It is therefore an interesting technology for the building industry that could contribute to energy-savings and improve the comfort level.

EXAMPLE: VACUUM GLAZING

Some developments have already been discussed in the previous chapter of this book. The following presents the example of vacuum glazing; new technology is developed to increase the thermal insulation and/or acoustical properties.

Vacuum glazing is a good example explaining the complex interdependencies

6 Three types of heat transfer through a cavity: conduction, convection and radiation

7 Relation between conduction, convection and radiation (their heat transfer coefficients α) in relation to depth cavity

of constructions and building physical performance:

As discussed earlier, one of the possibilities to improve the thermal insulating performance of double glazing is to evacuate the cavity or space between the two glass sheets. Recent years have witnessed in-depth research on this topic carried out by Professor R. E. Collins and his colleagues at the University of Sydney in Australia.[2, 3] Research on this topic has also been conducted at other institutions and companies, in Germany and the Netherlands, for example.[4]

Based on the research of Collins and his colleagues, the Japanese glass manufacturer Nippon Glass Co has developed the world's first commercialized vacuum glazing product. This vacuum glazing product, known as SPACIA, has been on the market since 1997. It is made of two 3mm thick panes with a 0.2mm vacuum space in-between, for a total thickness of only 6mm. Very small pillars in the cavity counteract the resulting pressure differences and vacuum-induced forces, making the glass structure sufficiently strong and rigid.

According to Nippon Glass, the thermal properties of SPACIA are approximately a quarter that of a 3mm-thick single pane and around half that of 12mm-thick double glazing. Due to its meagre thickness, vacuum glazing might therefore be especially suitable for installation into existing window frames of older homes and renovation projects. Despite the very slender glass pillars and the expected related acoustic leakages in the cavity, the manufacturer claims that SPACIA has also improved the acoustic properties. This is explained by the phenomenon that with conventional double glazing the sound transmittance by resonance between the panes causes less effective noise insulation than single panes with the same mass. With SPACIA glazing, the two vacuum panes are firmly pressed onto support pillars by atmospheric pressure, reducing resonance between the panes to a minimum, and thus resulting in a more effective sound insulation.

The conclusion that evacuated cavities can have a positive effect on sound insulation was also drawn from research carried out on vacuum glazing by the TU Delft as part of the ZAPPI project.[4] However, in this research it turned out that the design of spacers in the cavity – needed to achieve structural stability of the glass structure – resulted in acoustic leakages, with a significant negative influence on the acoustical properties.

CONCLUSIONS

In addition to their structural benefits, vacuum structures or building components show potential to improve the thermal and acoustic insulation performance of façade parts and the energy performance of the building as a whole. However, to use vacuum technology effectively in building structures in order to improve physical properties, one should be aware that – from a physical point a view – there are some restrictions and considerations regarding the proper use of vacuum.

Vacuum has a significant effect on thermal insulation only in cases of small cavity structures, or in case of structures with deep cavities if thermal loss due to heat radiation is minimized (use of E-coatings). It is also important to note that the level of vacuum has to be high enough; yet, these high states of vacuum mean stricter requirements on structural stability and aspects as airtightness, proper sealing, etc.

Considering acoustic aspects, the state of vacuum has to be relatively high to be effective. And to profit from the advantage of vacuum on acoustics, the amount of acoustic leakages need to be minimized.

Vacuum technology has great potential, but in order to solve these (practical) problems, and in order to use the vacuum technology effectively, more research and design improvements will have to done.

Arie Bergsma

LITERATURE
1 'Meerlagen model', ir. L. Nijs, Department Building Physics, Delft University of Technology
2 'Current status of the science and technology of vacuum glazing', R.E. Collins and T.M. Simko, School of Physics, University of Sydney
3 Caddet Energy efficiency: vacuum glazing with excellent heat insulation properties: "Spacia", (internet)
4 'Simulation of sound insulation properties of vacuum Zappi Façade Panels', M. van der Voorden, L. Nijs, H. Spoorenberg, Delft University of Technology

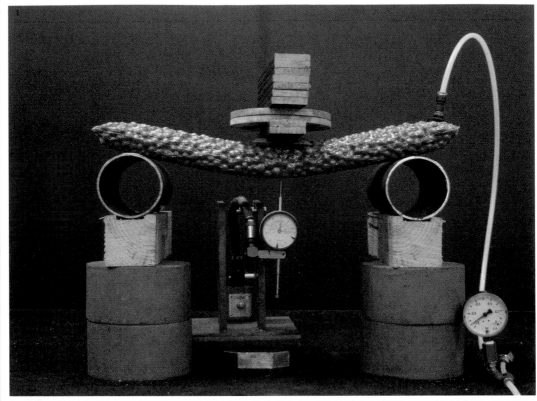

1 Vacuum coffee pack
2 Load test on vacuum beam, ETFE bag filled with
   hydro-granulate
3 Vacuum façade
4 All-round pressure on the balls caused by the
   achievable vacuum in the bag
5 Compression pre-stress of the balls

1.4. MECHANICAL BEHAVIOR OF VACUUM-SUPPORTED MEMBRANE STRUCTURES

1 INTRODUCTION AND GENERAL PRINCIPLES

The aim of this chapter is a general investigation of the mechanical behavior of vacuum-supported membrane struc-tures, which have the opposite structural behavior of air inflated structures.[1]

Structural concepts have bbeen developed to investigate and test the vacuum behavior as a load-bearing principle. One of these concepts comprises of a beam and an arch design, an example of which was built as a footbridge with a 10-meter span over the canal next to the building of the Faculty of Architecture of the Delft University of Technology (fig. 21). The idea for the vacuum footbridge is based on the principle of the "coffee pack" (fig. 2): loose aggregates, in this case hollow spheres (balls), in a closed membrane envelope (bag) pre-stressed by vacuum to form a rigid body with load-bearing capacity. Fig. 1 shows an example of a load test on an ETFE beam filled with hydro-granulates to illustrate the structural principle. This principle has been researched by load tests on 1:5 scale models.

The structural behavior of this type of vacuum structures is influenced by the material properties of the membrane (rate of elongation, strain), the rigidity (time dependent, creep) and the shape (which determines the degree of aggregate interlocking and influences the overall shape of the structure / membrane) of the aggregate particles, the practically achievable and maintainable state of vacuum, and the loading conditions a-symmetric loading).

There is also a scale factor between the size of the aggregate particles and the overall size of the structure (membrane) when determining the structural behavior. Further fundamental research is needed to determine this relation. The vacuum principle's potential for thermal and acoustic insulation (see chapter on the physical properties of vacuum) can also be use in claddings for façade systems (fig. 3).

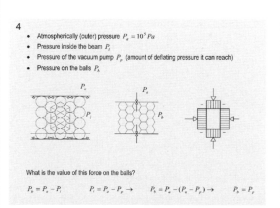

4

- Atmospherically (outer) pressure $P_a = 10^5 Pa$
- Pressure inside the beam P_i
- Pressure of the vacuum pump P_p (amount of deflating pressure it can reach)
- Pressure on the balls P_b

What is the value of this force on the balls?

$$P_b = P_a - P_i \qquad P_i = P_a - P_p \rightarrow \qquad P_b = P_a - (P_a - P_p) \rightarrow \qquad P_b = P_p$$

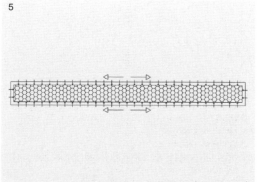

5

2 STRUCTURAL MECHANICAL VACUUM PRINCIPLES

The principle of vacuum is that by removing air from an enclosed volume the atmospheric air pressure will act as a load on the membrane surface. A perfect vacuum is impossible in practice; the acievvable state of vacuum depends on the capacity of the vacuum pump. Thus, the pressure on the loose aggregates / balls caused by drawing a vacuum with a vacuum pump is the achievable state of vacuum in the bag (see fig. 4).

By drawing a vacuum, an all-round compression force is exerted on the balls (fig. 5), perceivable as pre-stress. The tensile stress in the membrane can be determined on the basis of the equilibrium of forces (fig. 6).

3 LOAD TESTING OF VACUUM PRINCIPLE

The main issue when building a vacuum structure based on the "coffee pack" principle is stiffness, or lack thereof. It is therefore important to determine the modulus of elasticity of the composite material (bag with aggregates). Thus, axial load tests are done to determine the strain caused by the load. With these test results and the calculated stress, we can calculate the modulus of elasticity (fig. 7), which is time-dependent.

For the first test series to examine the load-bearing capacities of a beam based on the "coffee pack" principle (in this case on two supports with a uniformly distributed load, fig. 8, 9), a PES/PVC membrane bag was filled with loose aggregates (playpen balls, fig. 10) and the set under vacuum. The pre-stress caused by the vacuum results in compression stress, and the load applied results in bending stress in a cross section of the beam. As long as the compression stress can compensate the tensile stress caused by the load, the balls will remain under compression (fig. 11). If the tensile stress forces exceed the compression stress, the balls lose contact with each other and the tensile stress in the membrane increases.

When applying a vacuum, the ball configuration will be that of the closest possible packing of spheres. Because the balls used here are relatively soft, the high pressure of the vacuum creates maximum contact areas between the balls and a dense interlocking configuration; thus forming a relatively stable system of packing to be able to take the large pressure caused by the vacuum. However, the high compression stress under additional loading caused some balls to collapse (fig. 12, 13). Thus, the combination of relatively soft ball material and collapse of some balls both resulted in a low overall

---

6

Equilibrium is then:

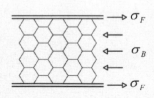

$$\sigma_B A_B = \sigma_F A_F$$

7

$$\sigma = E.\varepsilon \qquad \varepsilon = \frac{\Delta l}{l} \qquad \sigma = \frac{F}{A}$$

Δl = Length difference (reduction before and after use of force)
l = length
F = Force
A = Surface

The E-modulus can be derived to $E = \dfrac{\sigma}{\varepsilon} = \dfrac{F}{A} \cdot \dfrac{l}{\Delta l}$

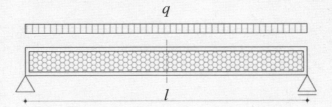

6 Tensile stress in the membrane
7 Axial test and formulas for determining the modulus of elasticity
8 The load-tested beam
9 The load-tested beam

11

Forces caused by deflating the beam (vacuum)   Moment caused by loading the beam

$$-P_b = -(P_a - (P_a - P_p))$$

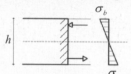

$$\sigma_o = \pm \frac{M}{W} \quad N = 0$$

$$\pm \frac{M}{W} - P_b = 0$$

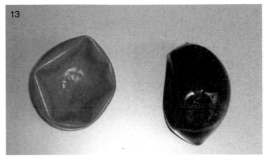

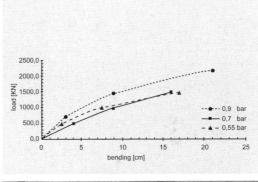

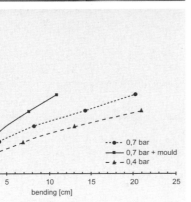

10 The playpen balls packed

11 The different types of stress in the beam

12 Collapsed ball caused by high vacuum pressure

13 Collapsed balls caused by high vacuum pressure

14 The load – deflection relation for the test with
   different states of vacuum with playpen balls

15 The load – deflection relation for the test with
   different states of vacuum with technical balls

16 Technical balls

stiffness of the beam. The test illustrated that the greater the vacuum in the bag, the stiffer the structure (fig 14).

Because of the poor performance of the playpen balls, "technical balls", usually used for water management (fig. 16), were tested. The tests showed improved performance; and once again a higher state of vacuum provided more stiffness (fig. 15). These balls will be used for further testing of structural systems.

4 STRUCTURAL MECHANICAL VACUUM SYSTEMS

The next structural vacuum system to be investigated is an arch with various loads (fig. 17). These loads represent the dead load (symmetrical loading) and variable loads, such as people and wind (non-symmetrical loading).

Because of the arching motion combined with the pre-stress caused by the vacuum, there will be more compression stress to compensate the tensile stress caused by bending moments as a result of the non-symmetrical loading (fig. 18).

One problem with non-symmetrical loading is that the hereby generated bending moments cause deformations, which, considering the low stiffness of the "vacuum bag principle", can result in large deflections (fig. 19). Such large deflections might lead to severe stability issues. It is therefore important to control these bending deformations. This can be accomplished by increasing the bending stiffness of the arch (increased height), or by other measures, such as cables along the arch functioning as a stiffening system based on the "Suchov principle" to reduce asymmetrical deformations.

Another aspect to consider are the forces applied to the arch support elements. For an arch to work as a system under compression, it is important that the support elements can carry the horizontal forces caused by the arch action. Large

stress forces are exerted onto the support elements which can cause large deformations. It is important to design the support in a manner that minimizes the stresses and deformations (fig. 20).

5 CONCLUSIONS

- Vacuum structures can be used as temporary structures, but rigidity properties need to be examined.
- To achieve structural rigidity of the entire vacuum structure, it is preferable to use aggregates with a high stiffness (thereby taking the time dependency of the modulus of elasticity, creep, into account).
- The properties (shape, size) of the aggregate particles play an important role in the structural behavior and require fundamental research.
- High performance vacuum-based structural systems can be built as long as the asymmetrical deformations caused by different loading conditions are controlled.

Andrew Bogaart

17

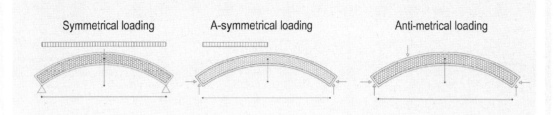

Symmetrical loading   A-symmetrical loading   Anti-metrical loading

18

Forces caused by deflating the beam (vacuum):

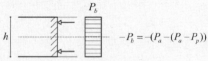

$-P_b = -(P_a - (P_a - P_p))$

Normal force caused by loading the beam:

$-\dfrac{N}{A}$

Moment caused by loading the beam:

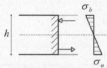

$\sigma_o = \pm\dfrac{M}{W}$

$-\dfrac{N}{A} \pm \dfrac{M}{W} - P_b = 0$

If we join these two and keep in notice that we want $|\sigma_o| \leq |P_b|$ we get the

19

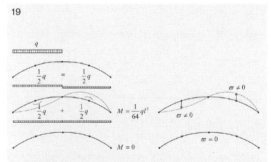

20

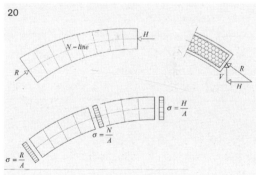

21

17 The arch with alternating loads
18 The different stresses in the arch
19 Asymmetric deformations as a result of bending by
   non-symmetric loading
20 Normal forces and stresses in the arch, and reaction
   forces on support elements
21 Vacuum arch

1.5. WORKING WITH DEFLATEABLES

This chapter looks at deflateables from a practical point of view and illustrates the experience we gained and the problems we ran into while working on various projects.

This chapter begins with discussing the materials required for deflateables, and the specific properties of vacuum constructions. Subsequently, the specific issues we encountered in applying the vacuum principle to different experimental projects are described.

1 WORKING WITH DEFLATEABLES IN GENERAL

Equipment

Working with vacuum requires special equipment, such as a vacuum pump, hoses, valves, and meters. The valves and meters work directly opposed to those for regular air systems. For use in vacuum applications a specific type of reinforced hose is used on order to prevent the hose from sealing itself off during negative pressure. Furthermore, we worked with a system of simple connections on all of the equipment, which allowed the rapid adaptation of the setup to varying projects and requirements.

A vacuum pump is required to create vacuum. For our purpose, a pump was required that was capable of running for long periods at a time, since we needed to deflate big volumes, as well as to maintain a vacuum, even when leakage occurred.

Another factor to consider is the capacity of the pump, since the time it takes to reach a certain pressure depends on this factor. For some projects it is useful to freeze the object rapidly, keeping it in the desired shape. The capacity of the pump also dictates the maximum achievable pressure difference which, in turn, determines the amount of force with which the filling can be pressed together.

Therefore, the pressure applied is an important factor in the structural use of vacuum. It is interesting to note that it is the force of the surrounding air pressing against the skin of the vacuum structure that applies the force.

Depending on the application, the pressure should suit the purpose. For the structural projects, we tried to reach the lowest possible pressure. With our semi-professional equipment we reached a maximum of -900 mbar. Another interesting observation is that the pressure will depend on the actual surrounding atmospheric pressure.

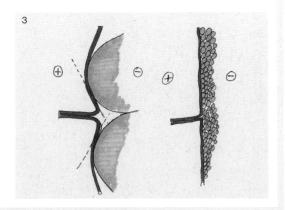

1 Vacuum meter
2 Paper models for patterning the ends of the enveloping foil
3 Smaller filling material creates a homogenous surface

The skin

In both of the principles a skin is used to encompass a filling or a frame. The skin comprises of a membrane or a foil. These foils have to exhibit a high-tensile strength and good airtightness.

The food industry has shown us that, if a structure is properly sealed, it is possible to maintain the vacuum state for a long time without having to continuously pump air out air. Of course adequate foil material and sealing are required to accomplish this.

Such gas-tight foils are also being developed for the use in vacuum insulated panels. The foils we used are more common in the building industry, such as PVC, PTFE and ETFE foils.

For several of our projects, we used a large square bag to create a straight beam or panel. But the use of membranes allows for all types of shapes to be used. A cutting pattern has to be designed for each shape. Here, the design of the seams is an important aspect; not only from an aesthetic point of view, but for manufacturing purposes and to accommodate stress forces in the foil. It can be useful to test the cutting pattern in a paper model (fig. 2).

When working on structural projects, it is important to consider the size and the shape of the filling material. With a bigger filling material, the gap between two individual pieces will be bigger; thus, the foil will be pressed in more in these places, introducing more stress. Therefore the tension on a foil with big filling material will be higher than that of a foil enveloping smaller filling material. Smaller filling material also creates a more homogeneous surface, see fig. 3.

These seams are made with an ultrasonic welding machine – which are large and expensive and therefore only used by larger companies – or by manual welding with a hot-air blower.

In order to reuse the envelope and change the filling material, we found different types of water and airtight zippers, scuh as thosee used for diving gear, for example.

Leakage

As with inflated structures, problems with leakage can occur in deflated structures.

This leakage can take place in 3 ways: due to the permeability of the skin, through the seams, and through punctures.

1 The skin

Similar to inflated pneumatic structures, there is always some diffusion of air through the membrane itself. This meant that with the foils we used, we needed to maintain the vacuum over longer periods of time.

For the structural projects, we used reinforced foils, because they can accommodate higher tensile forces; thus

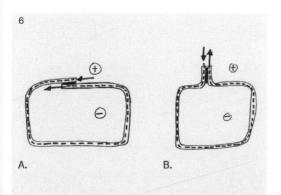

A.   B.

4 Welding a seam with a hot air blower
5 Welding a seam with a high-voltage welding machine
6 Preventing air diffusion along the textile re-enforcement

allowing for stronger structures. These foils consist of fiber-reinforced PVC. Here, extra attention has to be paid to the seam, because air might penetrate through the foil via the fibers (fig. 6). The airflow follows the fibers and is not hindered by PVC. With an overlapping seam, the fibers penetrate into the airtight compartment. Thus, the optimum airtight seam has both foil edges pointing outwards in parallel fashion, eliminating fibers that might penetrate the airtight compartment.

2 Seams

Compared with inflated structures, leakage via the seams is less problematic in deflateables. The negative pressure presses the two foils together, whereas positive pressure in an inflated structure tries to separate the foils. (fig. 7)

It was therefore possible to conduct tests on deflated beams without properly sealing the seam. To close the bags, we wrapped the end of the bag around a metal rod and clamped it together.

Vacuum squeezes the ends together providing a seam that is sufficiently tight to reach low enough pressures for testing. This simple method enabled us to empty and refill the bag rather easily. Thus, we were able to quickly test different materials.

3 Holes

In general, holes are dangerous for deflated constructions. However, the bigger the volume of the construction, the less critical leaks are, depending on the relation of incoming air versus the global volume. As long as the pump is capable of evacuating more air than the amount entering, the situation stabilizes after an initial significant pressure drop. Having a pump permanently attached can serve as a security function.

The same phenomenon occurs at the valve. Imagine an item getting stuck on the end of a vacuum cleaner, blocking the hose. The suction capacity diminishes. Therefore we welded a spacer into the valve in the bridge to prevent filling balls from shutting it of unintended (fig. 8). This ensured continuous suction.

Shrinkage

Specific properties of vacuum are shrinkage and creeping due to the negative pressure. The filling material reaches for its best and compact state of compression, resulting in shrinking of the structure. This effect depends on the form, size, and strength of the filling material as well as the friction between the particles and the amount of pressure applied. This process can continue after the deflation process and cause creeping, meaning a change of form due to further settlement of the

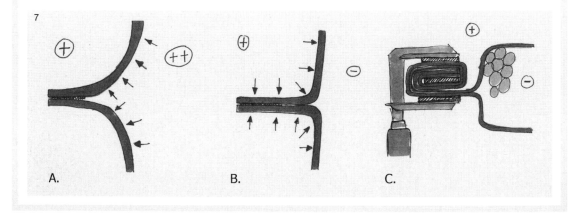

7

A.

B.

C.

7 Designing the seem
8 The distance keeper attached to the valve
9 The molded beam: The texture of the cardboard
   inlays is still visible
10 Cardboard panels in pockets to make a
   disappearing mould

See page 45
11 Sketch of the bag for the bridge, with cardboard
   panels in the middle
12 Holes for internal air-flow (FEDEX Facade)

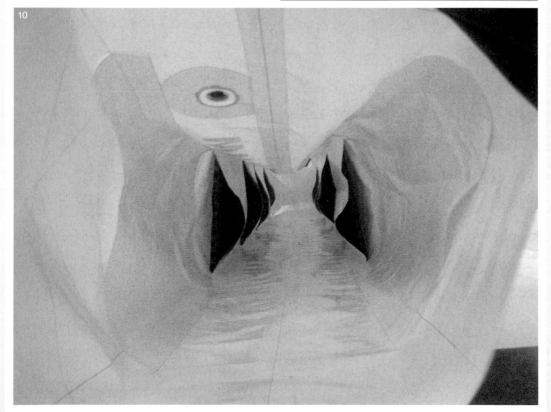

packing. The softer the filling material, the more compressed it will get, resulting in a weaker and rubber-like overall structure.

With the use of the harder filling materials (e.g. the harder balls in the bridge project), shrinkage will mostly be due to the material trying to find its ideal packed position and thus the least amount of space. During testing of the bridge project, we found that the shrinkage was approximately 6%. It is therefore important to take this shrinkage and a possible change of shape into account when designing the cutting pattern for structural projects, whereby accommo-dating potential change of shape is more difficult than estimated shrinkage.

2 PRACTICAL EXPERIENCE WITH THE STRUCTURAL PROJECTS

The structural application of deflated structures was extensively tested in a number of projects at the TU Delft. This culminated in building a deflated bridge. For this project an arch-shaped bridge spanning 10 meters was built. After testing different filling materials, plastic balls were used. These tests and the entire bridge project are discussed in other chapters of this book. In this context, I shall focus on some of the more practical problems we ran into while conducting the experiments and building the deflated bridge.

The skin used for this bridge was a long bag in the shape of a straight tube. The bag itself was not shaped in to an arch. This meant that the membrane would be more rippled on the underside of the bridge when formed into an arch. But, with this method the bag could be hung to find its own ideal arch shape. And it made the cutting pattern much easier, since it only required rectangular pieces. Testing showed that a rippled foil has minor influence on the stability of the construction until the bending stress

exceeds the pre-stressing forces of the vacuum; in this condition the system can deform, and the rippled, larger foil area cannot absorb additional loads. However, in this situation the deflated construction has reached its capacity.

Since the balls are all of the same size, they try to interlock in the optimum and densest pattern under the load of deflation. This pattern is predefined, and can be disturbed by a too-loose or too-dense filling. In addition the choice of same-size balls reduces free-formability. A certain mixture of differently sized balls could reduce these problems, but in practice, it would be very difficult to achieve a homogenous mixture for the entire structure.

All calculations were based on a square profile. Thus, in order to produce good results in testing, the beam needed to be square as well. Beams have the tendency to form into an oval shape before deflation. To prevent this from happening, experiments were done with molding the shape. As the vacuum sets in, the shape and position of the balls are set. A wooden mould was built to produce a square testing beam, which was squeezed into the mold and then deflated. This process worked quite well; resulting in a nice square beam. It also produced an improved surface structure, because the balls are held in a tight pattern.

However, building a mold is a lot of extra work. So the idea immersed of inserting the mold into the bag by sliding cardboard panels into pockets within the bag. (fig. 9-11) This kept the volume in shape until the deflation process started. We discovered that a light vacuum of about -100mbar is sufficient to maintain the shape. The cardboard is soft enough to structurally disappear under the full load of the vacuum. Therefore the cardboard does not influence the overall stiffness of the beam.

3 PRACTICAL EXPERIENCE WITH THE FAÇADE PROJECTS

We conducted several façade projects following the principle of inserting a frame into a membrane. When building these vacuum façades, the frame functions as a spacer to separate the two foils on either side. The negative pressure causes the two foils to try to adhere to one another.

Because the foil is pulled over the spacer with great force, it is important to avoid sharp edges on the spacers in order not to damage the foil. We used rounded frame edges as a precaution. In those places where the foil is pulled over the frame, peak stresses occur in the skin. The strength of the skin can be increased by adding reinforcement patches here places.

The spacers inside the foil need to accommodate the high pressure induced by the foil pressing against it. Therefore, the frame needs to possess some rigidity to prevent it from buckling. The stretch of the foil is an important factor. When using very flexible foil, the two layers will draw together extensively; eliminating any space in between. This diminishes the advantageous building physical properties of working with vacuum.

The distance between the transoms and mullions of the frame is another factor to consider. If the gap between them is too big, the foils will adhere to each other.

There needs to be a balance between the stretch of the foil and the shape of the frame.

It is important that the entire structure is deflated in a homogenous way. Because the foil is pulled tightly over the spacers, it might seal off individual compartments within the structure, even though they need to be connected. For the design of the of the FedEx façade, we incorporated holes in the spacing structure to allow free flow of the air inside (see fig. 12 and FEDEX Facade).

The design and fabrication of a deflated structure is a complex matter. We conducted a number of experiments to learn about the determining factors. The patterning of the cover has to be carefully designed with regards to airtight seam structures and the expected overall shrinkage of the overall structure. Hereby, the fabrication process has to be taken into account. The cover material itself has to be chosen according to the expected loads, which it has to transfer to the internal spacing structure. The shape and size of the spacing structure is responsible for the stresses in the cover material and the final capability of the structure.

Raymond van Sabben

11

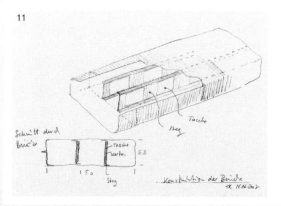

12

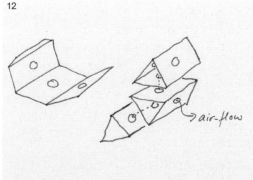

2. PROJECTS AND IDEAS

2.1. FAÇADES

The inherent force of air pressure can be used as an integrated element in the structural system of façades. Because of the additional properties such as sound and thermal insulation, a deflated construction has a high potential for the use in or as a façade.

This chapter focuses on systems with a light yet strong, spaced core which is subsequently stabilized by applying a deflated foil, glass or other airtight material. Considering these advantages, the aim is to conduct further research into vacuum façades with the combination of insulation properties and the possibility to serve as integral structural elements.

BALL-ENVELOPE
HONEYCOMB PNEU ENVELOPE
LE CADEAU, CUBIC PNEU WALL
MULTIFUNCTIONAL DEFLATED FAÇADE (REBUILDING TODS)
DEFLATED FAÇADE CONSTRUCTIONS 1
VACU-BAM
LOAD-BEARING EXHIBITION SKIN
IN-SITU VIP
BALL AND SOCKET
VACUUM ELEMENT FAÇADE
VACUUM SPACER
VACUUM COLLECTOR
DEFLATED FAÇADE CONSTRUCTIONS 2
BALLOON SUN-SHADING
BOOKSHELF FAÇADE
FEDEX FAÇADE
FAÇADES STRIPS

VISIONS FOR DEFLATED FAÇADES

Designing façades means to fulfill a range of principle functions; transparency being one of them. For deflated façade constructions it is necessary to apply the highest possible vacuum to achieve the best thermal acoustical insulation (see chapter 2.3. Building physical properties of vacuum structures). However, the forces caused by the air pressure pose a special challenge, if the need for transparency basically requires a dissolved core material.

The principles within this chapter imagine a range of possible solutions. They show systems or detailed solutions. Some try to make use of materials which are available onsite. A number of examples focuses on cardboard as a spacing material. Core structures made from cardboard can be transported to the site and allow easy assembly. They provide a lightweight and cheap solution which is easy to process. The deflated foil provides watertightness and stiffness in all directions. Cardboard is a good insulator and the dimension of the cardboard components can be designed depending on the specific structural needs.

BALL-ENVELOPE
05-09-2007

IMAGINED BY Tillmann Klein, Marcel Bilow, Harry Buskes
KEYWORDS freeform, pneumatic, system, insulation, load-bearing, lightness, structure, façade, envelope, structure, transport, 0-10 years, foil, membrane

Basically the construction consists out of a filling material and a foil to apply the pre-tensioning force. The ability of a filling to maintain its shape under the deflation pressure is important for strength and form precision (less creeping). Hollow plastic balls for example are strong, lightweight and translucent. The packing is essential for the free-form ability of the construction. Ball shaped elements in a single size tend to interlock into their ideal position with minimal volume, limiting freeform ability. A filling with multi-sized elements of course is harder to control in the mixture. A filling out of odd-shaped elements (like peanuts) will only be as strong as balls for the costs of more weight strength ratio. Packing can be controlled either by shaping the upper and lower foil into a quilt-like volume or by packing the filling material into pre-manufactured units. For the "Ball-Envelope" a strong and transparent foil is needed. Stretching of the foil and a soft filling will allow flexibility in the construction. Too much flexibility causes a lower e-modulus and has a negative influence on the stiffness of the construction. The strength is of course also limited by the air-pressure.

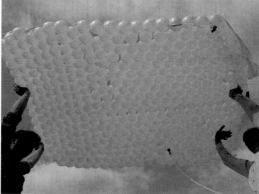

The filling can be controlled by packing it into pre-manufactured units. Connecting elements made of punched plastic sheets, known from can pallets, could be used.

HONEYCOMB PNEU ENVELOPE
22-02-2006

IMAGINED BY Jürgen Heinzel
SUPPORTED BY Marcel Bilow
KEYWORDS modular, system, ventilation, pneumatic, transparency, organic, façade, installations, foil, membranes, air

The solutions
- Modular frames that combines a set of frames to install like a element façade.
- Ventilation through a Gore-Tex Membrane, breathes but watertight.
- Moulded out of fibre-reinforced plastic.
- Good solution to fit the ceiling lining by adding mechanical service components like ventilation, cooling and also cable conduits.
- Innovative Idea for shading: in the chambers of the pneus, a can with a pneumatic balloon is filled into the pneu-chamber to regulate the shading.

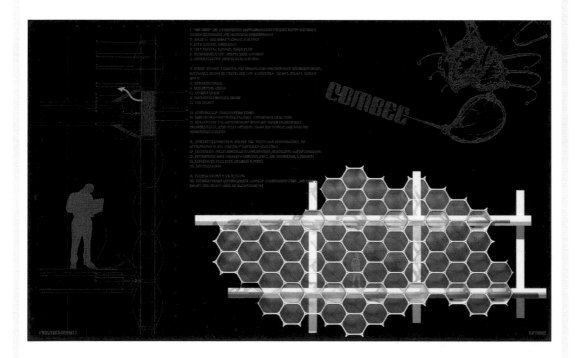

LE CADEAU, CUBIC PNEU WALL
13-03-2006

IMAGINED BY Luise Lauerbach, Thomas Hofberger
SUPPORTED BY Ulrich Knaack
KEYWORDS pneumatic, interactive, lightness, envelope, foil

This Idea is imagined, in a design project for a presentation system for the new Airbus A380. A lightweight presentation pavilion is build out of helium-filled cubicle Pneus, added on ropes like pearls. When you loosen the ropes, the wall starts to open by flying above the display object. A technical solution handles the control of the ropes in the construction shown at the bottom.

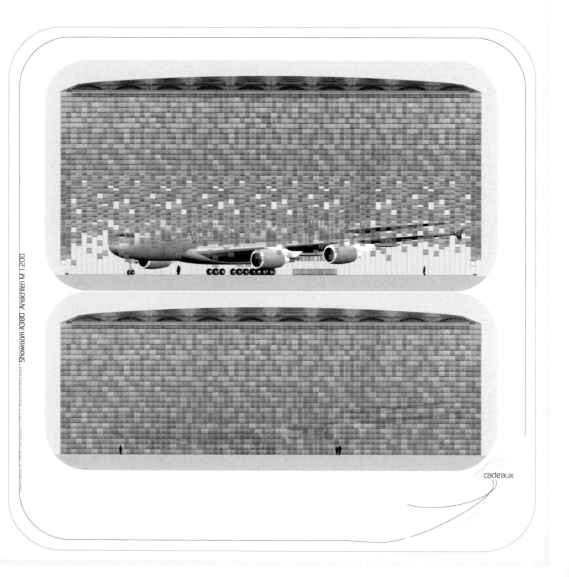

Showroom A380 Ansichten M 1:200

MULTIFUNCTIONAL DEFLATED FAÇADE (REBUILDING TODS)

22-05-2007

IMAGINED BY Tillmann Klein
KEYWORDS pneumatic, system, load-bearing, solid, façade, foil, unknown material

The façade consists of two quilted textile foils which apply pretension to the construction by deflating the spaces in between. The spacing material is filled in layer-by-layer. Different functions are represented by different areas with filling material of special properties (see below). Application of vacuum "freezes" the construction into shape. Even free-form constructions are possible. The whole façade can be reused and adapted, simply by refilling it according to a new chosen function-pattern.

A Structural areas, created by a strong filling on defined load paths
   inner zone: very dense material (thermal mass) / outer zone: lighter material (insulation)
B Translucent areas, created by strong, hollow, insulating spacer material
C Insulating area, created by compressible material. These areas are separately pressured in order to adapt insulation value (night cooling, etc.)
D Transparent areas with pressurized internal sun shading
E Inserted elements to allow façade openings
F Insulated infill frame to separate transparent areas (C)
G Air and watertight textile with varying surfaces, depending on covered areas

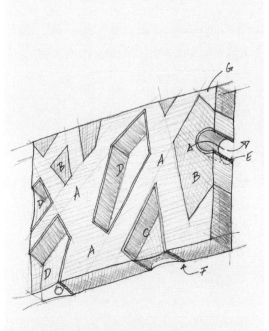

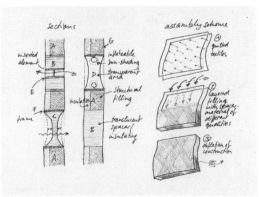

DEFLATED FAÇADE CONSTRUCTIONS 1
18-06-2006

IMAGINED BY Till Klein, Thiemo Ebbert
KEYWORDS free-form, pneumatic, system, mobile, load-bearing, lightness, adaptable, façade, envelop, structure, transport, 0-10 years, foil, membrane, aluminum

The benefit of evacuating an entire façade (panel) lies in the combination of stabilizing the structure by applying a defined pre-stress force with a high insulation potential. Space frame constructions are developed, adjustable in size and height.

Adjustable façade element
Flexible façade elements can be created when combined with vacuum skins. The construction can be easily transported because it takes up a minimum amount of space. The use can be temporary as well as permanent.

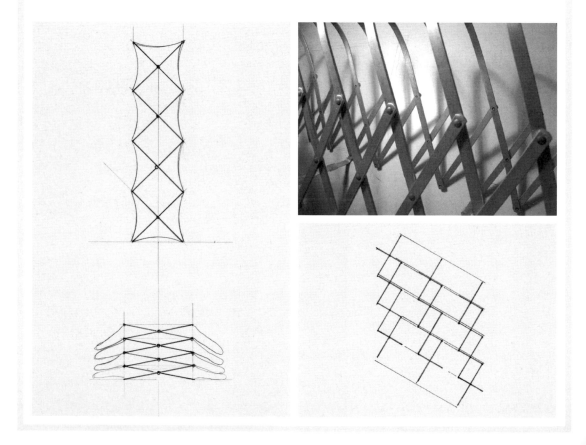

VACU-BAM
22-02-2006

IMAGINED BY Thiemo Ebbert
KEYWORDS composite, pneumatic, insulation, load-bearing, structure, foil, wood

A storey-high structure made of vertical bamboo columns forms the supporting structure. Two layers of transparent foil provide transparency and weatherproofing. By evacuating the space between the layers, the structure becomes highly insulated. Alternatively, timber poles or plastic tubes could be used instead of the bamboo columns.

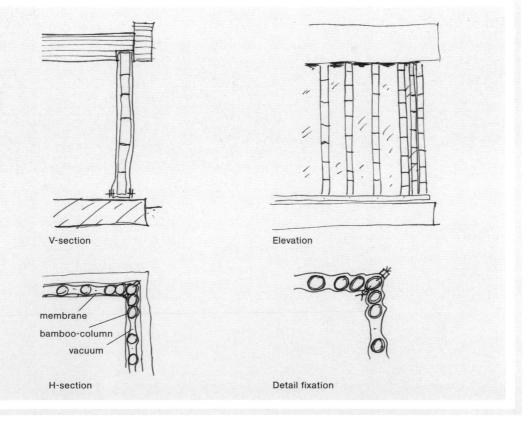

V-section

Elevation

membrane
bamboo-column
vacuum

H-section

Detail fixation

LOAD-BEARING EXHIBITION SKIN
27-03-2007

IMAGINED BY Thiemo Ebbert
KEYWORDS free-form, composite, pneumatic, load-bearing, commercial, low cost,
envelope, 0-10 years, foil

- Put all the stuff you want to show the world into a huge transparent plastic bag.
- Bring the bag into the desired shape suitable for your (exhibition) building.
- Extract the air.

The plastic foil contracts around the items inside. The items' dimensions provide the
necessary structural height. The pre-stress applied to the structure by the foil supports
the building.

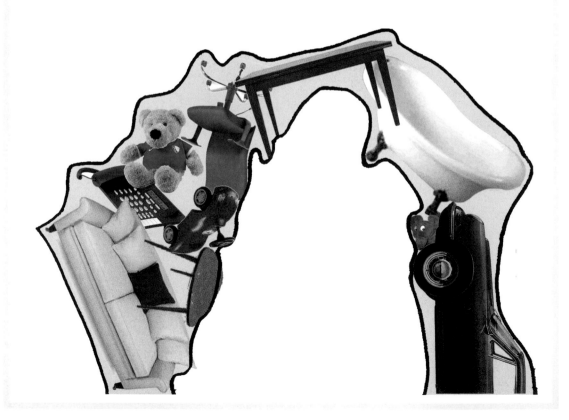

IN-SITU VIP
27-03-2007

IMAGINED BY Thiemo Ebbert
KEYWORDS on-site, free-form, composite, insulation, lightness, façade, building physics, 0-10 years, foil

Vacuum insulation panels provide highest insulation values at minimal thickness. Unfortunately, they are very difficult to produce in shapes other than flat panels. Thus connections, corners details and complex geometries are difficult or impossible to realize. The in-situ VIP comes as a cushion, filled with supporting material and air. It is positioned and shaped into place just like a blanket. By extracting the air the desired shape is reinforced and the high insulation level created.

Possible applications
- façade connections
- roller shutter housing
- free-form cladding

1
roughly dimensioned soft "VIP-blanket"

2
perfect fitting into shutter housing

3
evacuation of air provides highest insulation

BALL AND SOCKET
22-02-2006

IMAGINED BY Tillmann Klein, Marcel Bilow
KEYWORDS free-form, pneumatic, Insulation, transparency, lightweight, façade, envelope, structure, 0-10 years, foil, membrane

The aim is to create a lightweight, free-form structure with maximum transparency and a high insulation value. Here, a loose material forms into a 3-dimensional load-bearing structure when put in a transparent foil bag to which vacuum is applied. The construction works like a hard sponge and is here represented by ball-and-socket elements. Comparable with chemical reactions or magnetic systems, the ball automatically finds a socket to connect the loose ends. Depending on the chosen material and technology, the scale of the construction can range from millimeters to meters.

3-D ball & socket structure

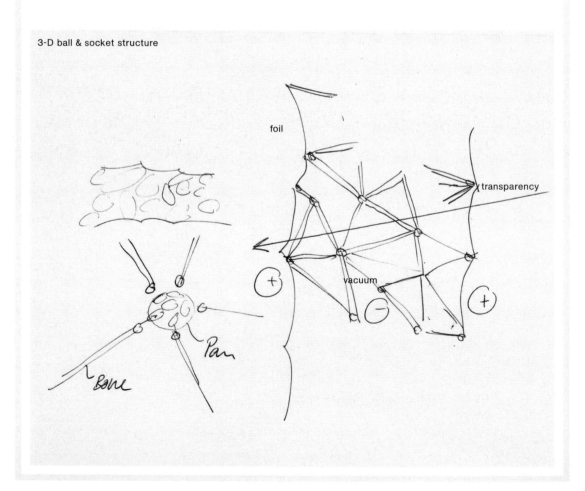

VACUUM ELEMENT FAÇADE
22-02-2006

IMAGINED BY Tillmann Klein
KEYWORDS pneumatic, system, insulation, load-bearing, transparency, low cost, lightweight, façade, envelope, building physics, 0-10 years, foil, membrane

This façade element is made out of lightweight frames covered with a foil. The frames act as a spacing element against the outside pressure. Part of their load-bearing capacity comes from the stability resulting from the vacuum-tensioned foil envelope. The vacuum is created by a pump that works with a tank. Each element is connected via a valve that prevents the system collapsing with the failure of one element. The elements can be combined into bigger structures, not only for walls but also for roof construction.
The construction combines transparency with high insulation value. Sun-shading can be inserted between the two foil layers in the vacuum chamber. It is a lightweight structure that can easily be transported and erected onsite.

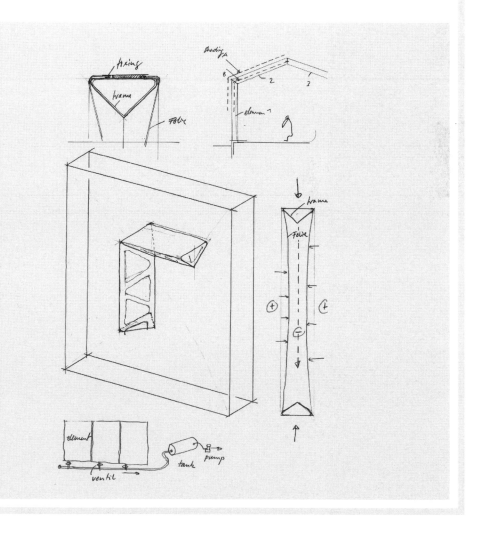

VACUUM SPACER
15-05-2006

IMAGINED BY Tillmann Klein
KEYWORDS pneumatic, system, load-bearing, lightness, transparency, façade, envelope, structure, building physics, 0-10 years, foil, plastics, air

Vacuum constructions need spacing elements. This spacer is designed to keep the two sides of foil covered vacuum chamber apart. It is made out of PTFE and based on the principles of water bottles. The relative low-pressure in the system will create an overpressure within the spacer. The circular cross-section increases the load-bearing capacity. The transparent spacer can be optimized by using alternative materials.

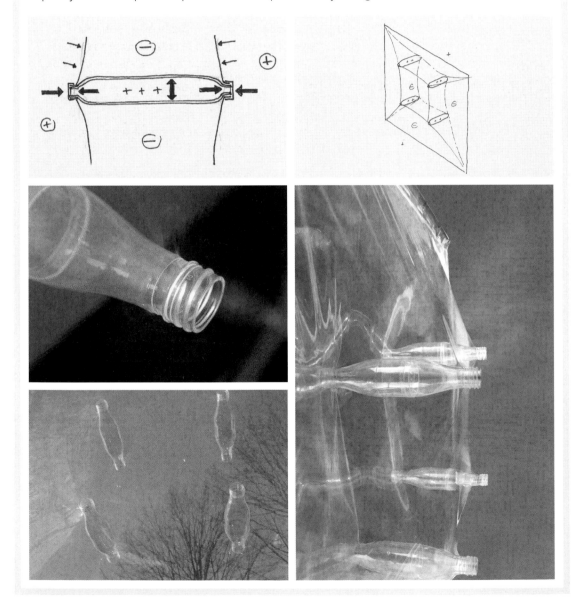

VACUUM COLLECTOR
22-02-2006

IMAGINED BY Tillmann Klein, Marcel Bilow
KEYWORDS free-form, pneumatic, system, energy generating, insulation, sun shading, lightweight, façade, envelope, installations, building physics, 0-10 years, foil, plastics

The Vacuum Collector is especially made for the vacuum-element façade and ideal as an application for greenhouses. A capillary hose is mounted on a dark cloth. This construction works like a sunshade and can be lowered to provide total shading for the room.
The solar energy is exploited by constant water flow inside the hose, powered by the capillary force or the upward movement of heated water. The effect can be enhanced by employing small pumps. The cloth will prevent heat radiating to the inside of the house and the vacuum exhibits a high insulation value. Advantage can be taken from the large water reservoirs that greenhouses need. The water is heated up. Inside the tank, warm water collects at the top, and cold water stays at the bottom. The form of the tank influences this effect. When needed, the soil around the roots of the plants can be heated or cooled through pipes that work like a floor heating system – common use for greenhouses. The combination of collector, vacuum element, water storage, and the mass of the soil creates a system than can be run reciprocally to regulate the climate.

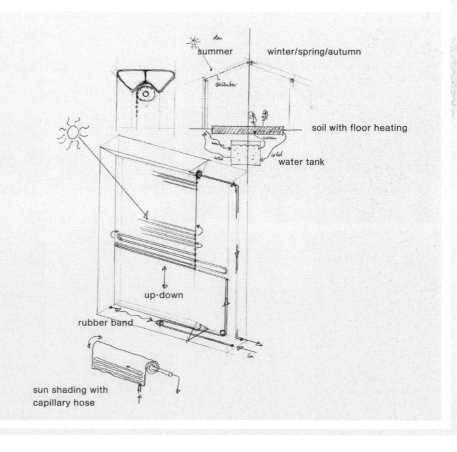

DEFLATED FAÇADE CONSTRUCTIONS 2
18-06-2006

IMAGINED BY Lourdes Lopez Garrido AR0645
KEYWORDS pneumatic, system, insulation, lightness, transparency, structure, façade, envelope, transport, building physics, 0-10 years, foil, membrane

The benefit of evacuating an entire façade (panel) lies in the combination of stabilizing the structure by applying a defined pre-stress force with a high insulation potential.

"Honeycomb panel" façade
Deflated constructions need spacing elements. Honeycomb cardboard has great stability along the direction of the holes. If an outside frame is provided, honeycombs can be used as filling for the panels in combination with a deflated foil. The result is good insulation value, light weight, and transparency. Alternatively the structure can be combined with a frame that can be disassembled.

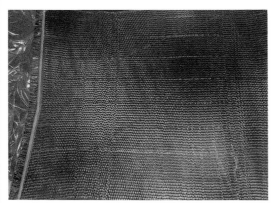

BALLOON SUN-SHADING
13-03-2006

IMAGINED BY Marcel Bilow, Tillmann Klein
KEYWORDS pneumatic, sunshading, moving, adapting, façade, envelope, 0-10 years, foil, air

The combination of deflated (façade) and inflated constructions (e.g. balloons) can be used to create sun-shading devices or different transparencies in façade constructions. Generally there are two possibilities to adjust the balloons.

1 **Open System:** The balloons are connected to an independent pressure system. The relation of pressure difference between inner and the surrounding pressure.

2 **Closed System:** The balloons are not connected to the outside pressure. By simply changing the vacuum within the façade construction, a pressure difference is created. The balloons have to be filled with a certain amount of air before being enclosed in the façade in order to allow a growing and shrinking of the device.

The First option is constructively more complex, but allows better regulation of the size without changing the vacuum in the façade and its insulating properties. Tests have shown that both principles work. A critical factor is the homogeneous pressure distribution within the system. Balloons attached in a cardboard bookshelf façade before the foil is mounted.

BOOKSHELF FAÇADE
05-09-2007

IMAGINED BY Lourdes Lopez-Garrido, Tillmann Klein, Robert Barnstone, Arie Bergsma, Raymond van Sabben
KEYWORDS free-form, pneumatic, system, insulation, load-bearing, lightness, structure, façade, envelope, transportation, 0-10 years, foil, membrane

The construction is based on a grid-structure made of cardboard. The 2-dimensional elements are pre-cut and pre-shaped and then assembled onsite into bookshelf-like 3-dimensional façade structures. The construction can be combined with sun shading devices. (See also: "Balloon Sun shading")

FEDEX FAÇADE
05-09-2007

IMAGINED BY Tillmann Klein, Robert Barnstone, Arie Bergsma, Raymond van Sabben
KEYWORDS free-form, pneumatic, system, insulation, load-bearing, lightness, structure, façade, envelope, structure, transport, 0-10 years, foil, membrane

The construction is based on triangular cardboard tubes, known from the FedEx poster shipping boxes. It is an inexpensive mass-product that forms the grid for a 3-dimensional façade structure. Transparent openings can be created by leaving tubes out. The construction forms a pressure envelope around the cardboard elements.

In order to create extra stiffness, the tubes are filled with triangular cardboard pieces. Enhanced transparency can be created by using stronger tubes.

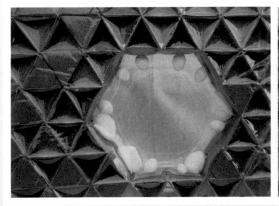

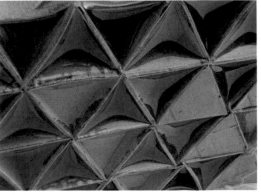

FAÇADES STRIPS
05-09-2007

IMAGINED BY Robert Barnstone, Tillmann Klein, Arie Bergsma, Raymond van Sabben
KEYWORDS freeform, pneumatic, system, insulation, loadbearing, lightness, structure, façade, envelope, structure, transport, 0-10 years, foil, membrane

Prefabricated cardboard strips form a structural mesh. The size of the structure can vary according to the desired strength and transparency. With help of the interlocking design of the strips, this construction provides great rigidity in the direction of the façade layer itself and, at the same time, stability when the vacuum pre-tensioning is not applied.

DEFLATEABLES IN STUDENT FAÇADE DESIGN

In the Master's program for Building Technology, the first semester focuses on research based design of an office building façade. Titled "Component and System Design", this course is based on a timetable within which the students develop a design from a concept or "façade scenario" to the final design. The first three weeks of the twenty-week curriculum are dedicated to the façade scenario, an innovative concept for the façade of a given office building. The best fifty per cent of the scenarios are selected by the teachers to be elaborated by teams of two students.

Modeling of this scenario is one of the most important methods in design research. For sound and thermal insulation computer modeling, programs such as VABI and TRISCO are available to the students. DIANA is used to model the mechanical behavior of the "scenario" design. With modeling software such as Pro Engineer and Rhino, virtual prototyping is included in the CAD part of the curriculum. In addition to computer modeling, material prototyping of the scenario design is a very important part of the curriculum. Intensive instruction in machining, welding and sheet metal handling equips the students for the material prototyping of their façade design. The education of a technical designer is not complete without hands-on building experience of a design. Material properties and production techniques are as critical for the end result as the design concept itself.

PROTOTYPING DEFLATEABLES

For the semester of September 2006, the assignment was complemented with a new constraint. Stimulating technical students to conceive innovative façade scenarios had proven to be difficult. Thus, new methods to force students to be as innovative as possible had to be found, resulting in the topic of vacuum or deflating, given within the Chair of Design of Constructions headed by Prof. Knaack.

The need to achieve a high level of vacuum for optimum thermal and acoustic insulation (see also the chapter on "Building physical properties") results in extreme loads on the structure. Under practical conditions, a vacuum of -850mbar delivers $850KN/m^2$. The forces within a façade created by such a vacuum are about one hundred times greater than the usual forces on a facade construction. Considering the trend toward lightness and transparency in façade design, this was a special challenge in the assignment for this semester.

Educationally speaking, nothing is wrong with a difficult mission in as much as failure is one of the most effective ways of learning. Material prototyping is one of the most honest and efficient ways because a collapsing part of a prototype, for example, is obvious to everyone. To see hidden problems in drawings is much more difficult and less obvious. For a headstrong design student it is much easier to keep believing that his or her design is okay when the critique is based on drawings rather than a real material prototype.

SCENARIOS FOR DEFLATEABLES

As mentioned above, the students of this first semester Component and Systems Design had to present an innovative concept for a façade of an office building within three weeks. The main challenge the students face is real innovation.

In case of the semester of September 2006, the additional assignment made the use of vacuum obligatory without the requirement to specify the exact technological method to be used.

A deflated cavity was, of course, used in the majority of the presented scenarios as the obvious way to meet the challenge of the assignment. The main problem within this solution is to avoid physical contact between the outer and inner sheet of the façade. The task to create a façade with an excellent sound and thermal insulation motivated the students to choose the deflated cavity as the main part of the design.

Four scenarios were eventually turned into material prototypes, and just one of these prototypes was able to withstand the pressure of the vacuum; a double sheet of glass with colored spacers. To improve the remaining three designs to prevent collapse under vacuum pressure is a difficult, but certainly not an impossible task. Material properties and production techniques are the starting point for improvement.

CONCLUSION

Educationally, material prototyping of the deflated façades was a success, even if the majority of the prototypes failed the final test with only one exception. As mentioned before, failure is one of best ways to learn.

To push the students to real innovation in façade design was the main goal for including the deflation constraint into the assignment for that particular semester. Innovative design will bring students to the edge of research by modeling and at the same time supports independent thinking. All of the prototypes of the scenarios can indeed be labeled innovative.

Peter van Swieten

DEFLATED GLASS HEXAHEDRON FAÇADE
REED FAÇADE
VARI-VIP
SUSPENDED CAVITY

DEFLATED GLASS HEXAHEDRON FAÇADE
20-01-2007

IMAGINED BY Jan Willem Hennink in collaboration with Edward Reijnders
SUPPORTED BY Peter van Swieten
KEYWORDS pneumatic, system, insulation, transparency, façade, structure, glass, steel

A double façade with a deflated cavity consisting of glass sided hexahedrons combined with tetrahedrons and a steel frame. The deflated cavity provides a strong thermal and sound insulation. Bending moments in the plane glass sheets requires rather thick glass layers. A dome-shaped structure would have been more efficient, but would result in much higher production costs.

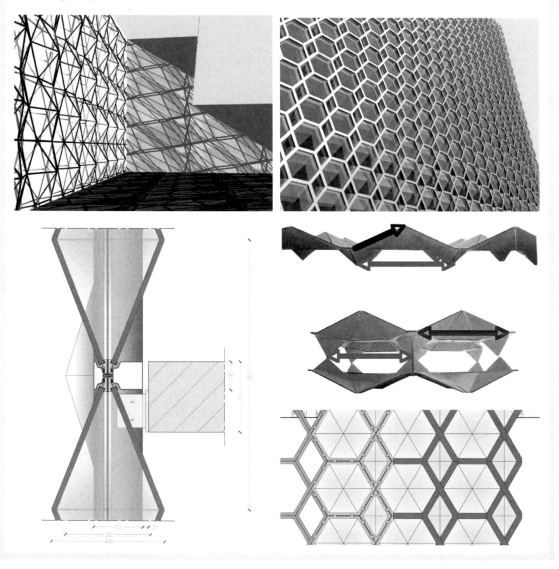

REED FAÇADE
20-01-2007

IMAGINED BY Tim Beuker in collaboration with Juan Davila
SUPPORTED BY Peter van Swieten
KEYWORDS pneumatic, system, insulation, transparency, façade, structure, membrane

The main material used for this façade is the sustainable material reed. It can be tied together in any desired form. This form is fixed by the application of vacuum, which, at the same time, increases the insulation value. Windows are created by voids in the filling. The picture in the lower left corner shows an example where the surface of the façade is shaped to allow the optimal airflow for wind generators attached to the edge of a building.

VARI-VIP
20-12-2007

IMAGINED BY Rosalie van Dijk, Janneke van Kilsdon
SUPPORTED BY Peter van Swieten
KEYWORDS pneumatic, system, insulation, transparency, façade, structure, glass

VARI-VIP is a double glazed floor-to-floor façade with a deflated cavity and the colorful use of spacers. By changing the distance between the spacers and their material, the façade can be adapted to different conditions or demands. Most of the research in the material prototype focused on the detailing of the edge of the glass sheets. The type of glue that can be used to withstand the air pressure is an important factor, and requires further research.

SUSPENDED CAVITY
20-12-2007

IMAGINED BY Niels Eilander, Sofia Candenas, Edward Reijnders
SUPPORTED BY Peter van Swieten
KEYWORDS pneumatic, system, insulation, transparency, façade, structure, foil, steel

This scenario avoids the use of spacers by introducing an outside construction to absorb the tension created by the vacuum in the cavity. Thermal leakage through internal spacers can be avoided. Point loads on the membrane need to be evenly distributed, which could be done by using eyelets similar to those of a sail. The external construction looks rather bulky; a cable suspension system could provide a lighter solution.

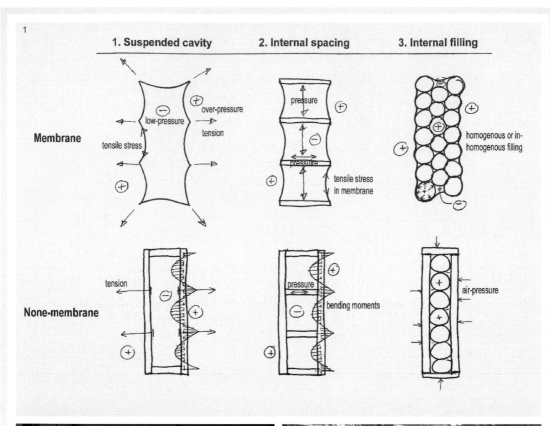

1. Suspended cavity   **2. Internal spacing**   **3. Internal filling**

Membrane

None-membrane

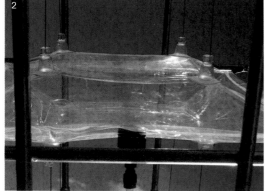

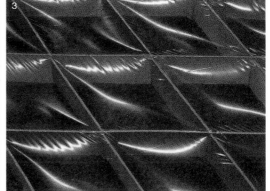

1 Different types of deflated structures
2 Suspended cavity
3 Three-dimensional spacer made of two-dimensional
   cardboard strips

2.2. STRUCTURES

Basically, two components are needed for this sort of structure; an airtight enveloping component and a spacing component. The enveloping material functions as the pre-tensioning element, whereas the spacing material creates the necessary structural height and distributes the pressure loads. Vacuum tries to compress the structure into its smallest possible volume. Thus, in order to create a shape, a certain strength of the spacing material is needed to overcome this tendency. Generally, it must be lightweight, yet strong, to minimize the effects of shrinking and creeping; factors that have to be taken into account in all deflated constructions. The spacing component can be a packed material such as the coffee in coffee packs or it can be a special structure of linear, two or three-dimensional components. The enveloping material can be a membrane or a non-membrane able to handle bending moments.

There are three different ways of creating a deflated structure (see fig. 1):

1 THE DEFLATED CAVITY
A deflated cavity is created by suspending the enveloping material with an outside structure (see example fig. 2). The key factor in designing suspended cavities lies in the connection between the outside structure and the enveloping material. This can be done in a linear way or with selective spot fixtures, leading to stress peaks in the envelope. An advantage of this structural type is that there is no direct connection between the inner and outer side of the envelope, thus thermal and acoustical leaks can be prevented.

2 THREE-DIMENSIONAL INTERIOR SPACERS
Three-dimensional interior spacers can be used to stabilize the enveloping material from the inside of the cavity (see fig. 3). All kinds of constructions are thinkable. If linear spacers, such as PET bottles (see fig. 4), or two-dimensional spacers are used, a support in the third dimension might be necessary, since the air pressure impacts the structure from all sides. That could be accomplished by an outside structure or a frame. Here, the connection between spacer and enveloping material is also critical, but the advantage in comparison to the suspended cavity is that the the entire construction does not exceed the size of the desired cavity.

3 INTERNAL SPACING WITH A HOMOGENEOUS OR NON-HOMOGENEOUS FILLING

A filling material can be used to distribute the pressure loads (see fig. 5). The properties of the filling, the weight, its e-modulus and long term structural behavior have a direct effect on the behavior of the entire system. The form will influence the packing, friction between the elements and the load distribution in the system. The size of the filling particles in relation to the size of the construction matters. A packing consisting of one single type of round elements causes the filling to interlock in a straight pattern. Different sizes and shapes can generate a denser and more shapeable packing; preferable for the use in free-form structures. Peanuts with their different shapes and rough surface, for example, displayed good properties for free-form constructions.

Generally, the combination of covering material, spacing material and pressure determines the final strength. Depending on the spacing, a deflated construction might creep during and after the deflation process, a factor that needs to be considered.

Pressurized spacers in the deflated cavity can lead to lightweight structures. A non-membrane as an enveloping material (e.g. glass panes, fig. 6) can provide some of the spacing functions due to the internal resistance to the bending stresses from the air pressure. Depending on the pressure differences and distance of spacers, this can result in quite large material dimensions (glass sizes), making this type of construction potentially more heavy than deflated membrane structures.

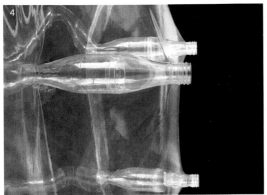

4 Linear "pet-bottle" spacer
5 Internal spacing with inhomogeneous material
  (peanuts)
6 Glass-sheets as enveloping material with acrylic glass
  spacers

DEFLATED BRIDGE
FREE CURVED VACUUM PANELS
ZIPPER
PET BOTTLE STRUCTURES
PET BOTTLE CUBE
DEFLATEABLE JOINTS 1
DEFLATEABLE JOINTS 2
DEFLATEABLE JOINTS 3
DEFLATEABLE JOINTS 4
VACUUM DOME
ARTIFICIAL FILLING MATERIAL
DEFLATED BUNDLE BEAM

DEFLATED FREE-FORM STRUCTURES

Depending on the filling of deflated structures, the structure can be shaped into any desired form before the deflation. By evacuating the air, the form is then "frozen". The spacing material interlocks through the pressure applied, so that the shape of the construction becomes fixed. According to this principle, temporary building envelopes or lightweight load-bearing structures can be created that can be adapted to the requirements of the respective situation.

The stiffness of the filling material, the friction and the strength of the covering membrane are essential for the final load-bearing capacity of the structure and its shrinking and creeping behavior during and after the deflation process.
Our main focus is on temporary uses, since the potentially higher pressure of deflated vs. inflated structures can cause problems that are not yet solved. Leakage can result in structural failure and backup constructions lead to more complex and expensive solutions.
New technologies in membrane materials such as strength, transparency and assembly properties will provide further opportunities for this constructional principle.

DEFLATED BRIDGE
18-06-2006

IMAGINED BY Marcel Bilow, Tillmann Klein, Thiemo Ebbert
ELABORATED BY Wouter Blondeel
KEYWORDS free-form, pneumatic, system, load-bearing, lightness, structure, envelope, 0-10 years, foil, membrane

"Peanut" bridge

There are different possibilities to build a deflated bridge. One of them is the peanut solution. A material is packed into a pre-shaped bag. The ideal arch form can be identified in a hanging position. After evacuating the air, the construction is stiff and can be turned upside down into its correct position. "Peanuts" represent all positive characteristics a filling material must exhibit: optimum shape for good packing properties, friction due to its rough surface, lightweight. This principle was tested in a 10-meter-long bridge, with a filling material of hollow plastic balls, since peanuts were not readily available in the enormous quantity needed. We do not want to propose this experimental structure for actual bridge constructions. However, it is a great example of testing and verifying assumptions and calculations.

This project was sponsored by: Carpro.be / Eynatten Bubbledeck / Leiden

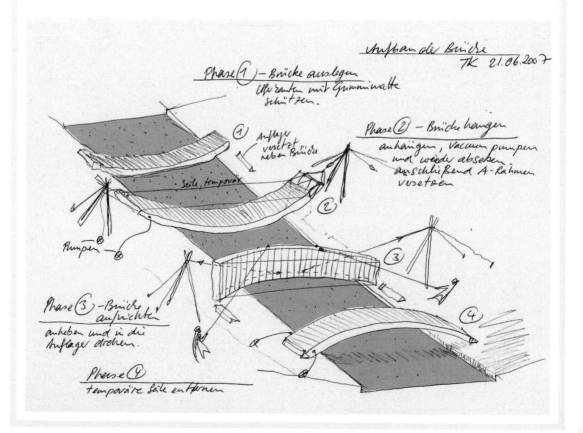

FREE CURVED VACUUM PANELS
15-06-2006

IMAGINED BY Ulrich Knaack, Daan Rietbergen
KEYWORDS layered construction, free-form, composite, load-bearing, structure

The concept behind this idea is to create curved panels by using intelligent cutting patterns and vacuum. The sandwich panels are made of a soft (bendable) surface with a stronger core. Lines are milled into the core-material along according to the desired folding pattern. The core is then wrapped in foil and deflated using a vacuum pump. During the deflation, the sandwich structure folds along the predefined, milled lines. More research needs to be done to calculate the cutting patterns for freely double curved surfaces.

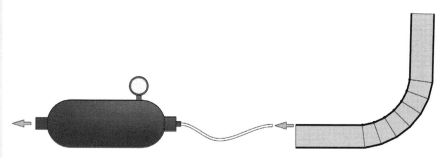

Lines are cut in core material

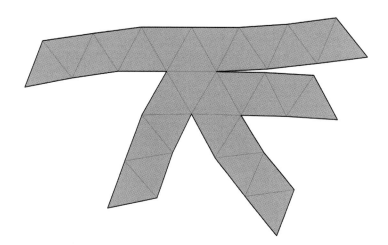

In order to create double curved panels, appropriate cutting patters need to be researched

ZIPPER
14-08-2007

IMAGINED BY Tillmann Klein, Marcel Bilow
KEYWORDS pneumatic, other function, strength, structure, foil, air

In order to reseal deflated constructions airtight zippers can be used. Zippers are flexible and can bear loads up to 300N/cm perpendicular to the zip-joint; thus accommodating surface stresses that the foil is exposed to.

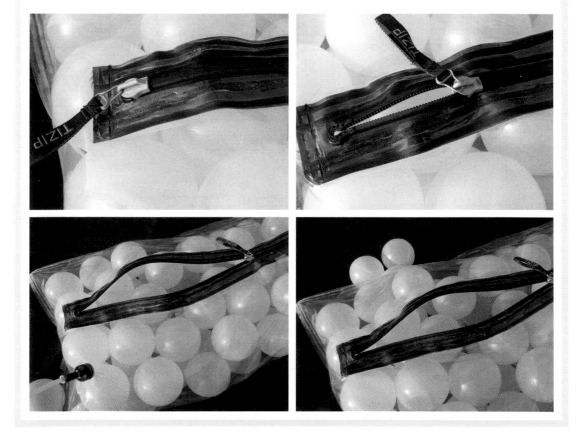

PET BOTTLE STRUCTURES
29-08-2007

IMAGINED BY Ulrich Knaack, Marcel Bilow
KEYWORDS pneumatic, load-bearing, transparency, structure, plastics

When searching for alternative filling materials for deflated structures, PET bottles are a cheap and transparent filling material. Research and tests have shown that it is possible to erect self-supporting structures using this material. The bottles themselves are not as stiff as hard plastic balls; thus, structures using PET bottles are not able to support additional loads, but they can support their own weight when used for walls or temporary structures. One of the advantages of using PET bottles is their transparent to translucent appearance, allowing illuminated, shiny structures for interior or trade show applications.

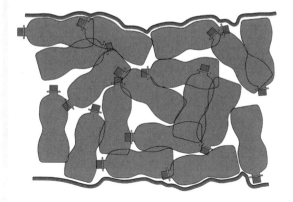

PET BOTTLE CUBE
29-08-2007

IMAGINED BY Marcel Bilow
KEYWORDS pneumatic, load-bearing, transparency, structure, plastics
DESIGN / CONCEPT OF BOOTH Display International
CLIENT Plastic Europe

This project was designed as a trade show pavilion made with PET bottles. The embedded steel structure takes the load of the second level, which is used as a VIP room, and also accommodates the vacuum pipes. Thus, the load-bearing functions and the technical equipment needed to erect the stand are integrated in the steel structure. Integrated lights illuminate the pavilion during show hours – creating a special look and serving as advertisement media.

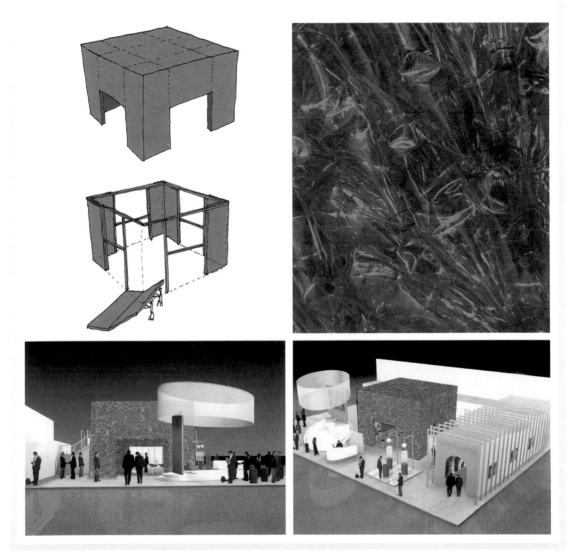

DEFLATEABLE JOINTS 1
20-12-2007

IMAGINED BY Tillmann Klein
KEYWORDS pneumatic, load-bearing, strength, structure, unknown material

Using vacuum might provide a new solution to fixate joints. This concept illustrates a joint that is able to fixate occurring shear forces. The vacuum connection created in a small airtight chamber is necessary only to link the elements together. This joint can be used in systematic or modular building elements that require minimum time for erection or disassembly, or are used in temporary constructions. The concept eliminates the need for small hardware components such as screws or bolts; thus also eliminating the possibility of losing these small items.

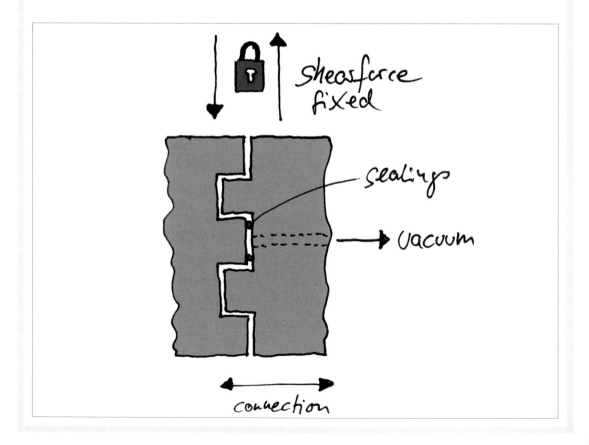

DEFLATEABLE JOINTS 2
20-12-2007

IMAGINED BY Tillmann Klein
KEYWORDS pneumatic, load-bearing, strength, structure, unknown material

This is another project using vacuum to fixate joints. This concept illustrates a joint that can lock components together and also provides an active sealing method. The vacuum is established in a circulated chamber. This chamber is surrounded by rubber seals that secure the vacuum and also keep the entire joint wind and waterproof. This type of connection provides durability; due to the permanent low-pressure it adjusts itself and reacts on the possible movement of the connected elements. A control unit measuring the pressure could detect air leakage. If the vacuum pressure runs below a predefined minimum value, a warning signal could be triggered.

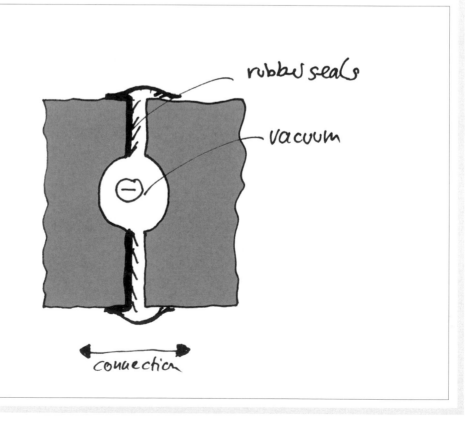

DEFLATEABLE JOINTS 3
20-12-2007

IMAGINED BY Tillmann Klein
KEYWORDS pneumatic, load-bearing, strength, structure, unknown material

The main idea behind this concept is a pre-compressed linear sealing that can expand after assembly. A foam or similar expandable material is compressed into an airtight skin or envelope and is attached to the joint element in a conformal, key-like profile geometry. After placing the elements together, the skin opens and the foam expands into the defined open space; thus creating an interlock that fixate the elements. The process can also be reversed to take the elements apart when needed.

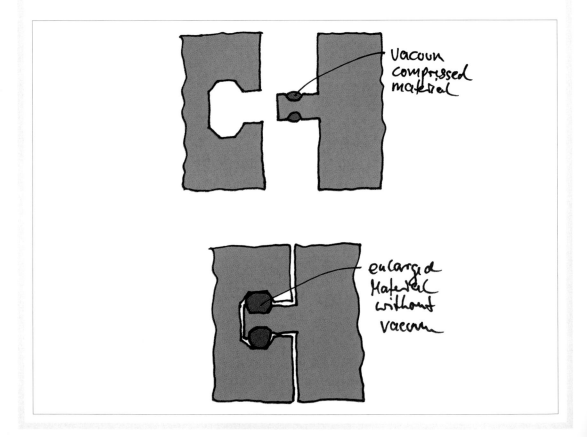

DEFLATEABLE JOINTS 4
20-12-2007

IMAGINED BY Tillmann Klein
KEYWORDS pneumatic, load-bearing, strength, structure, unknown material

Using vacuum might be a solution for fixing joints. Hydraulic or pneumatic stamps are well known for movement application in all areas of industries. Inverting the principle of overpressure into low-pressure a length adaptable element is possible that enables a tractive force to interlock elements. These elements could be used as additional elements in combination with standard linear sealing joints that provide an inner tractive force at the intersections of building elements to enhance the performance of the sealing joints.

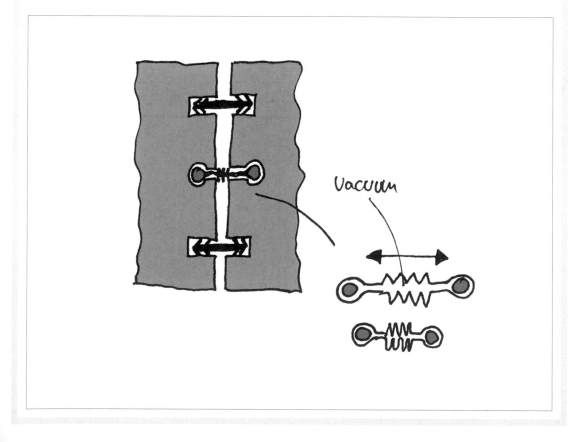

Vacuum

VACUUM DOME
29-08-2007

IMAGINED BY Ulrich Knaack, Marcel Bilow
KEYWORDS pneumatic, load-bearing, transparency, structure, plastics

This project shows a dome structure to illustrate the load-bearing potential of deflated constructions. The filling material is a combination of plastic balls and PET bottles. The hard plastic balls are used in areas that require greater stiffness, and PET bottles in other areas to allow for illumination and transparency.

ARTIFICIAL FILLING MATERIAL
29-08-2007

IMAGINED BY Marcel Bilow
KEYWORDS free-form, pneumatic, load-bearing, strength, structure, plastics

In general, the load-bearing capacity of deflated constructions (the coffee pack principle) depends on the filling material. Promising test results with peanuts as filling material stimulate more research into finding optimum filling materials with a higher load-bearing potential. The peanuts' shape and surface texture exhibit very good packing and interlocking properties of the filling. Especially the free-form capability of the structure is great; however, it does not provide sufficient stiffness for most structures. The artificial peanuts shown here were designed via CAD and manufactured with a 3-D printer to test the geometric and structural behavior. These filling materials could easily be made in large quantities by using plastic injection molding.

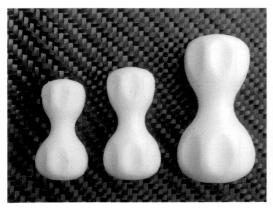

DEFLATED BUNDLE BEAM
02-03-2008

IMAGINED BY Tillmann Klein, Marcel Bilow
KEYWORDS free-form, load-bearing, adaptable, structure, membrane, unknown material

A free-form beam can be created by bundling flexible strands, enveloped by an airtight membrane. Three jacketed control cables are imbedded in the outer perimeter of the bundle, used to preshape the structure. The change of shape causes the strands to shift. Deflation of the structure increases the friction between the strands, giving the bundle strength. A special end-unit is needed to compensate the shift of the strands and provide a flush finish.

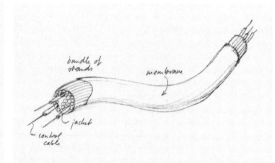

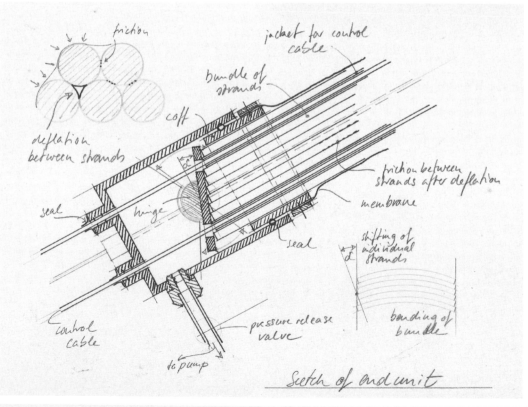

FOLDING A SANDWICH BOAT
SELF-ERECTING CONSTRUCTION
KINETIC VACUUM STRUCTURES

DEFLATED KINETIC STRUCTURES

If preformed elements are inserted into an airtight bag and vacuum is applied, the structure can mould into this predefined shape and has an implanted "memory", which can be temporarily erased by re-inflating the structure.

This principle can also be used like a muscle, meaning that the form can change continuously with by varying the air pressure; thus, moving structures are also imaginable. The filling material plays an essential role. First trials with sandwich constructions as predefined foldables and structures representing the "muscle effect" have been successfully conducted to test their basic functionality.

FOLDING A SANDWICH BOAT
28-06-2006

IMAGINED BY Herbert Funke, Marcel Bilow
KEYWORDS layered construction, load-bearing, low-cost, structure, composites

A glass-fiber-reinforced plastic (GRP) boat was developed during a plastic seminar.
The sandwich core, providing structural height, serves as a lost mold: thus, no additional
molding forms are needed. The boat is constructed in 4 stages: first, the foam core is laid
out and coated with GRP on one side. Pressure is then applied by a deflated, enveloping,
airtight plastic bag. When the coating is cured, this plane element is turned around, and
the foam core is cut along the folding pattern. Tests showed that the outer GRP skin does
not delaminate during the folding. Finally, the inner side is also coated with GRP, keeping
the folded form in shape and providing the final stiffness.

This project was supported by the following companies: Gebr. Becker GmbH & Co KG / Wuppertal (vacuum
pump), Hyco-Vakuumtechnik GmbH / Krailling, R&G GmbH Faserverbundstoffe / Waldenbuch (GRP), epurex
films / Walsrode (foil for vacuum pressure)

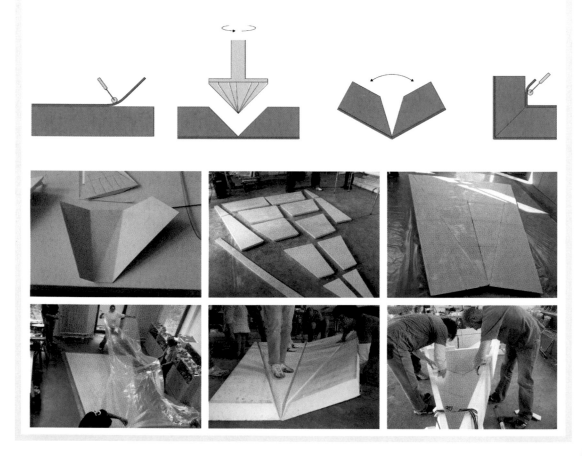

SELF-ERECTING CONSTRUCTION
15-06-2006

IMAGINED BY Marcel Bilow
KEYWORDS pneumatic, mobile, load-bearing, adaptable, structure, foil, composites

Self-erecting constructions can be created by inserting precut elements into an airtight bag and applying vacuum pressure. This type of 'biform' construction exhibits two shapes; one in its non-deflated and a second in the deflated state. For easy handling and transportation, the non-deflated elements need less space than they do in their erected final position. The kinetic principle applies to this construction as it "grows" or moves into its final shape.

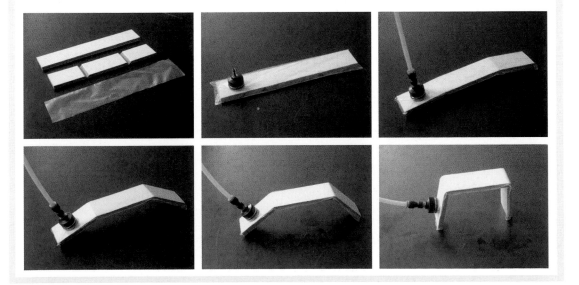

KINETIC VACUUM STRUCTURES
28-06-2006

IMAGINED BY Marcel Bilow
KEYWORDS pneumatic, mobile, adaptable, structure, tool, 0-10 years, textile, paper, composites

Inserting a rollable structure into an airtight bag enhances the kinetic behavior.
The cardboard used in this project is rolled in under vacuum and rolled out when inflated.
The phenomenon determines that any structure under vacuum morphs into its smallest dimension and shape. This technique can be used for packaging and covering, as well as for structural use due to the high stiffness of the evacuated core.

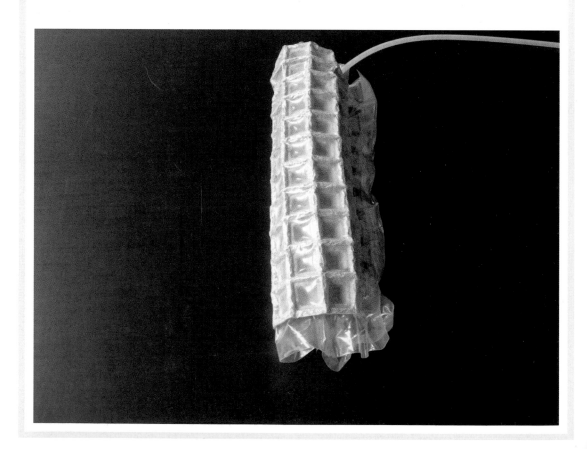

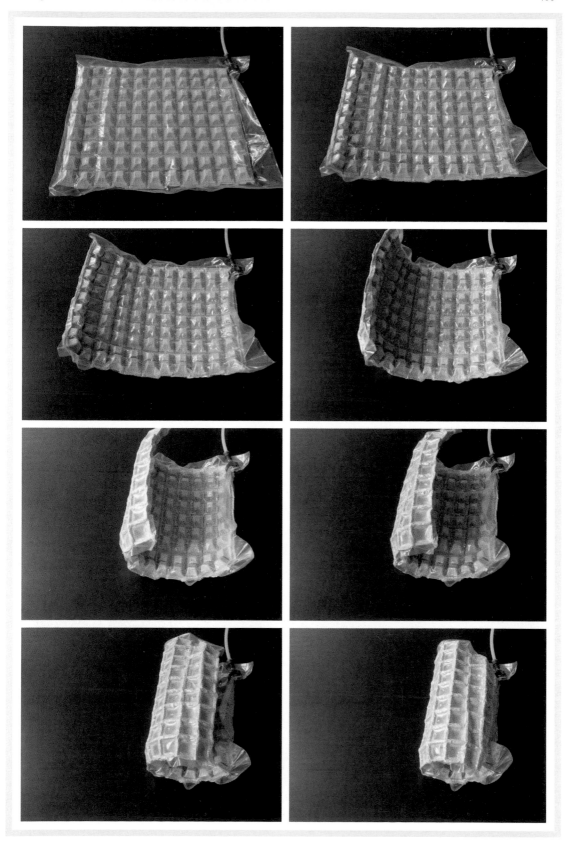

AIRIK
AIR-B-WALL
SELF-ERECTING CHAIR
FLEXCURVE
LOUNGE CHAIR
INFLATABLE SHOWER
CUBE-X
QUILTING TECHNIQUES FOR DEFLATED FURNITURE
MATTRESS

2.3. FURNITURE AND INTERIORS

The concept of deflation can be particularly useful for the construction of furniture. Deflateable furniture could be inflated, re-shaped and than re-deflated if a change of shape is desired. Different surface textures can be created by different filling materials and the hardness can be changed by a change of the vacuum level in the system.

Another advantage that should not be under estimated is that a possible leakage might not cause more damage than what happens when an air mattress fails. The principles of this chapter show some general application possibilities and manufacturing details.

AIRIK
14-08-2006

IMAGINED BY Julia Schulte
SUPPORTED BY Ulrich Knaack, Marcel Bilow
KEYWORDS self organizing, pneumatic, mobile, commercial, interactive, system, foil, air

This project is a presentation robot unit with a chassis with wheels and a robotic control unit that scans the surrounding area, and lets the unit move around corners and obstacles. Presentations or images can be projected on to the upper surface of an inflated air envelope. The Airik can be used as a presentation system at fairs or other public events.

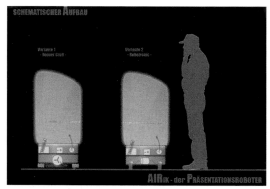

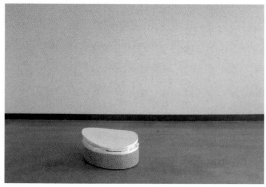

AIR-B-WALL
14-08-2006

IMAGINED BY Jürgen Heinzel
SUPPORTED BY Ulrich Knaack, Marcel Bilow
KEYWORDS free-form, pneumatic, mobile, adaptable, façade, structure, adjustable mold, foil, air

This project is based on using inflated balloons inside a flexible, airtight bag to create an adjustable wall.

These are a few ideas following this concept:
1　All balloons inflated to create a rectangular wall after deflation.
2　Lines of inflated balloons to create a moving wall.
3　Fill each balloon separately to achieve the most individual shape.
4　The airflow/suction could be regulated by a computer controlled valve system to let the wall move when different pressures are applied.

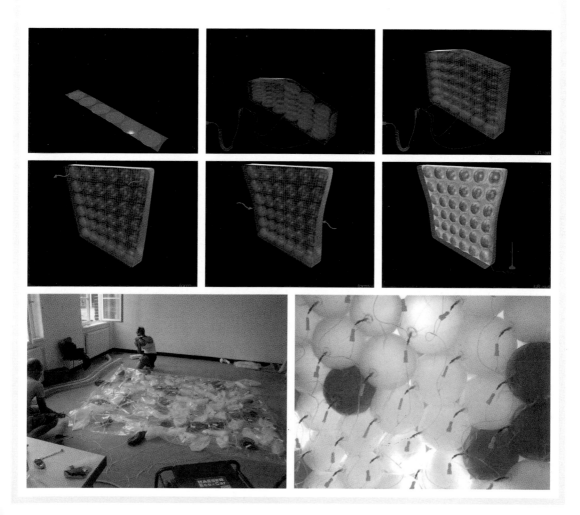

SELF-ERECTING CHAIR
07-11-2007

IMAGINED BY Daniel Neumann, Christopher Bahlke, Fabian Riemenschneider, Nikolai Aldinger
SUPPORTED BY Tillmann Klein, Marcel Bilow, Holger Techen
KEYWORDS free-form, pneumatic, other function, adaptable, foil

Based on the principle of mobile folding structures, this idea illustrates the functionality of a self-erecting chair. The shape of the non-deflated chair makes it very easy to transport. When in place, the chair folds into the desired form during deflation and exhibits sufficient load-bearing capacity.

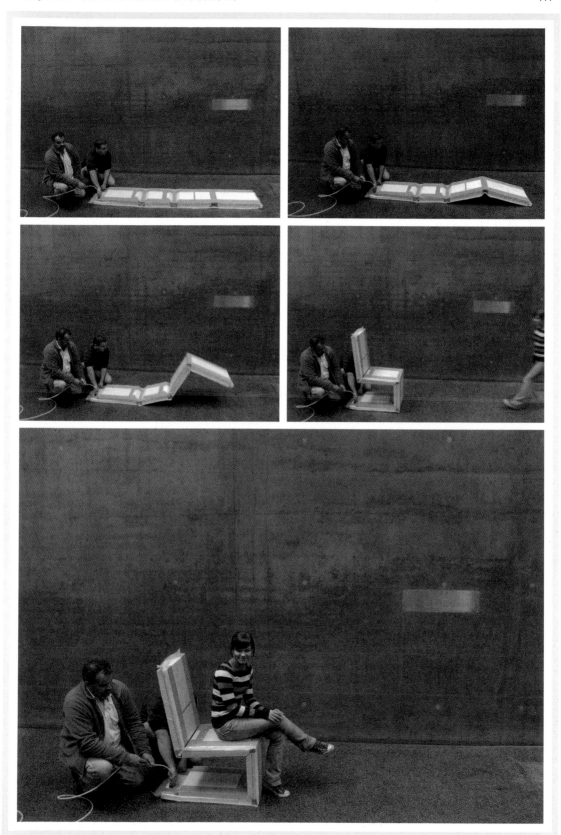

FLEXCURVE
07-11-2007

IMAGINED BY Tillmann Klein
KEYWORDS free-form, load-bearing, adaptable, structure, plastic

This idea is based on the principle of the flexible curve template. The template uses a flexible outer contour profile and a flexible inner core to create an inner friction, which helps to maintain the desired shape. When applied on a larger scale, this principle can support the shaping process and provide temporary rigidity to a structure before it gets "frozen" into its final shape under vacuum.

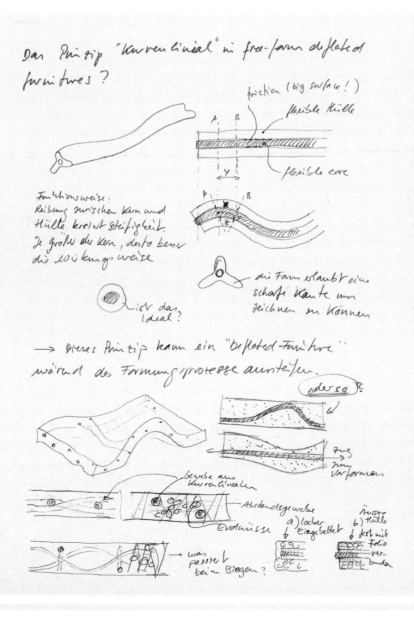

LOUNGE CHAIR
10-08-2007

IMAGINED BY Patrick Böhm, Roland Cubius, Konstantinos Efremidis
SUPPORTED BY Tillmann Klein, Marcel Bilow, Holger Techen
KEYWORDS free-form, pneumatic, other function, adaptable, foil

Free-form furniture can be created by using deflated structures. They can be easily adjusted by the user and serve various functionalities. The same object could change from a sofa into a table. The furniture is made of an air-tight membrane in combination with a filling, following the so called "coffee pack" principle. The foil determines the surface quality, and has to be strong enough to withstand the air pressure. The filling is responsible for the packing, strength and formability of the furniture.

We imagine the following fillings:
1. Colored balls, also known as IKEA balls. They are soft and make a dense packing after deflation. Tests have shown that the strength of the final structure is limited, but sufficient for furniture.
2. Hollow plastic balls. They are rigid and provide good strength. For enhanced free-formability, the packing has to include a mixture of differently sized balls.
3. Resin-coated peanuts (coated for increased strength).
4. Artificial peanuts, produced by rapid manufacturing. This process guarantees a desired strength value and a controllable mixture.

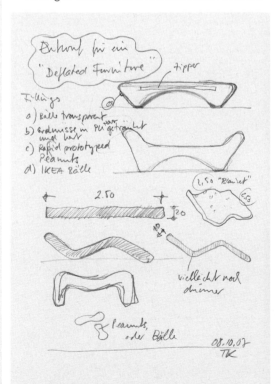

INFLATABLE SHOWER
14-08-2006

IMAGINED BY Mathias Kemper, Jan Erik Gerdt, Janko Grode
SUPPORTED BY Ulrich Knaack, Marcel Bilow
KEYWORDS free-form, pneumatic, load-bearing, commercial, system, foil, air

The idea behind this project was to create a self-supporting, mobile shower cabin by producing an envelope with shaped foil strips attached to different air chambers. The banana-shaped chambers are attached to each other on the outer perimeter of the construction. Since the inflated chambers push toward the outside, the outer skin is under tension which increases the overall strength of the construction.

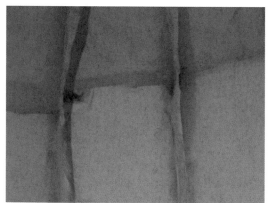

CUBE-X
14-08-2006

IMAGINED BY Dennis Schlepper, Konstantin Reimer
SUPPORTED BY Ulrich Knaack, Marcel Bilow
KEYWORDS free-form, pneumatic, mobile, adaptable, façade, structure, system building, foil, air

Different types of walls or display stands can be built by creating modular air-filled boxes with a plug system. The system could also be used as temporary partition walls, to protect certain areas from others, for example. The plug system can be realized by using air-filled tubes with valves, similar to bicycle tubes. Earplug-like foam plugs can be used to connect the elements.

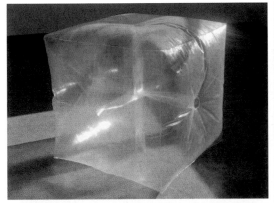

QUILTING TECHNIQUES FOR DEFLATED FURNITURE
10-08-2007

IMAGINED BY Tillmann Klein, Marcel Bilow
KEYWORDS free-form, pneumatic, other function, adaptable, 0-10 years, foil, air

The filling is a basic structural component of deflated, free-formable furniture. The moving and shifting of this filling has to be controlled. A certain amount of shifting is necessary to allow the furniture to change into its final shape; however, it needs to be limited in order to prevent the furniture from loosing this shape.

These are possible quilting techniques:
1 Pockets to keep the filling from moving excessively. This can be accomplished with perforated membranes or mesh material to maintain an undisturbed airflow in the system.
2 Cables connecting the upper and lower membranes are arranged on a pre-defined grid to keep the thickness of the structure within its specified limits. Alternatively, the two membranes can be connected by tubes. These tubes are open to the outside and can be used to help forming the furniture before deflation.
3 The furniture is filled with a flexible structure, such as a 3-D mesh made of soft fibers. This keeps the desired filling in place, but allows it to move during the deflation process.

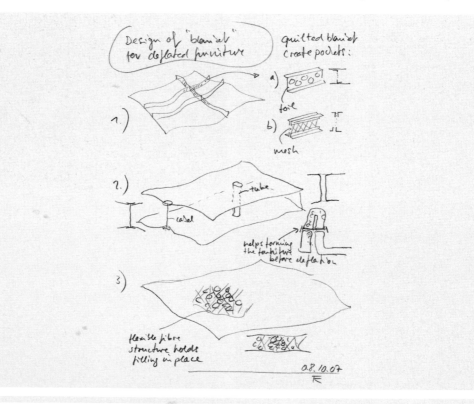

MATTRESS
10-08-2007

IMAGINED BY Felix Hartz, Frank Herzog, Mareike Strohbach
SUPPORTED BY Tillmann Klein, Marcel Bilow, Holger Techen
KEYWORDS free-form, pneumatic, other function, adaptable, foil

The simplest idea of a vacuum supported structure is an airtight bag and a filling material. These simple seats or mattresses remind us of the bean bags of the 1970s, but can be "frozen" in the desired form. Obviously, the comfort depends on the level of vacuum applied and the size and shape as well as the degree of softness of the filling material.

3. PERSPECTIVE

PERSPECTIVE

Sustainability and energy performance are some of the most important factors in the building market today. Here, deflated constructions bear high potential: minimal use of material, high insulation values, and a positive effect on transportation cost and logistics.

STRUCTURE

Various types of deflated structural systems have been researched. In general, an enveloping and airtight covering material and a spacing component are needed to create a deflated structure. Pre-stress can be used for stability by applying low pressure. The pre-stressing force cannot exceed the atmospheric pressure and is therefore limited. The combination of covering material, spacing material and pressure determines the final strength. Depending on the properties of the spacing component, a deflated construction might creep during and after the deflation process, a factor that needs to be considered. The enveloping material as well as the packing material, if used as a spacing, are also responsible for the final surface texture. The entire system has to be design for its specific purpose. For some of the examples, intensive research has been conducted, showing that the structural aspects of deflateables are a highly complex matter.

INSULATION

Besides its potential in a structural sense, the possibilities of vacuum as a method of insulation are vast, but this application has restrictions. The size of the cavity has to be taken into account and the vacuum level has to be high enough to achieve proper results. Thermal and acoustical leakages at the system's boundaries must be prevented. However, the potential for good insulation is a particularly important advantage that deflated structures have over inflated structures.

CONSTRUCTION AND DESIGN

As a rule of thumb, high vacuum levels provide the best results, but that also means high forces within the structure – a challenging factor, especially for transparent constructions with a minimized internal spacing structure. The loads from the atmospheric pressure have to be distributed via the enveloping material. Glass panels have to be significantly thicker and foils must resist the stresses without stretching or creeping. Existing material exhibiting sufficient strength are reinforced, and therefore do not yet provide the desired transparency. A high level of airtightness is needed, resulting in an increased effort when it comes to sealing and detailing. Redundancy is an important factor for the design of deflateables, since the

available materials and sealing techniques are still limited. The problems
are the same as those with the widely applied pneumatic structures.
If a structure relies entirely on the pre-stressing forces of vacuum, leakage
causes serious risk. Possible solutions are the installation of backup
systems or using vacuum only for minor supporting members. An example
would be to attach different functional layers of a façade system to achieve
the best performance. A failure of the system would cause the layers to
detach, but would not risk the collapse of the entire façade. It is also
possible to use lower pressure levels, resulting in simpler detail solutions.
Leakages are less dangerous and can be temporarily backed up by an
over-dimensioned pump.

THE FUTURE OF DEFLATEABLES

In the short term, we predict a focus on applications such as temporary
structures, furniture or packing materials. Specifically the possibility to
"freeze" free-forms into their final or a temporary shape offers a big range
of future applications.
In the long term, we expect a wider use for vacuum supported constructions, since technological possibilities will increase. Its application in
façades seems promising.
The examples in this book illustrate many possible shapes and ornaments,
which are truly derived from bare structural needs and can be adopted by
architectural design.

Further research into the field of packing materials and the effect on
structural and formative behavior, sealing and jointing methods and
membrane technology, in particular, would be beneficial for future
developments.
The combination of deflated systems with other structural principles such
as inflated systems or other means of pre-stressing could significantly
enhance the possibilities for future applications.

In concluding this book, we would like to point out that, despite or perhaps
because of the wealth of potential ideas and concepts, many restrictions
still apply to deflateables structures and many problems remain to be
solved. This fact is an integral part of an imaginary approach – identifying
current boundaries is as much a result of good research as new concepts
and technologies.
We hope our approach has encouraged the reader, and provides motivation
to pick up ideas from this collection and carry on discovering deflateables.

APPENDIX

CVs

ULRICH KNAACK (*1964) was trained as an architect at the RWTH Aachen, where he subsequently obtained his PhD in the field of structural use of glass. In subsequent years, he worked as an architect and general planner with RKW Architektur und Städtebau, Düsseldorf, winning several national and international competitions. His projects include high-rise buildings and stadiums. Today, he is Professor for Design and Building Technology at the Delft University of Technology, Netherlands, where he established the Façade Research Group and is also responsible for the Industrial Building Education research unit. He organized inter-disclplinary design workshops such as the High-rise XXL. Knaack is also Professor for Design and Construction at the Detmolder Schule für Architektur und Innenarchitektur, Germany, and author of several well-known reference books.

TILLMANN KLEIN (*1967) studied architecture at the RWTH Aachen, completing his studies in 1994. He subsequently worked in several architectural offices; from 1996 onward he was employed by Gödde Architekten, focusing on the construction of metal and glass façades and glass roofs. At the same time, he attended the Kunstakademie in Düsseldorf, Klasse Baukunst, completing the studies in 2000 with the title "Meisterschüler". In 1999, he was co-founder of the architectural office rheinflügel baukunst with a focus on art-related projects. His practical work includes the design of a mobile museum for the Kunsthaus Zug, Switzerland, the design and construction of the façades for the ComIn Business Centre, Essen, project management for the construction of the Alanus Kunsthochschule, Bonn, project management for the extension of the University of Applied Sciences, Detmold. In 2005, he taught building construction at the Alanus Kunsthoch-schule, Bonn-Alfter. The same year, he was awarded the art prize of Nordrhein-Westphalen for young artists. Since September 2005 he has led the Façade Research Group at the TU Delft, Faculty of Architecture.

MARCEL BILOW (*1976) studied architecture at the University of Applied Science in Detmold, completing his studies with with honors in 2004. During this time, he also worked in several architectural offices, focusing on competitions and later on façade planning. Simultaneously, he and Fabian Rabsch founded the "raum204" architectural office. After graduating, he worked as a docent and became leader of research and development at the Chair for Design and Constructions at the FH Lippe & Höxter in Detmold under the supervision of Prof. Dr Ulrich Knaack. Since 2005, he has been member of the Façade Research Group at the TU Delft, Faculty of Architecture.

ARIE BERGSMA (*1971) studied aerospace engineering at Delft University of Technology. After graduation in 1995, he worked as materials researcher at Hoogovens R&D, Product Application Centre (now TATA steel). From 1998 until 2004, he studied architecture and building technology at Eindhoven University. Before and during this period of part-time study, he worked at several engineering offices in the Netherlands: Prince Cladding BV, D3BN Structural Engineers and Peutz Consulting Engineers. At this last office, he worked as a consultant on building physics and acoustics from 2001 until 2006, and was involved in several large-scale building projects in the Netherlands (head offices of Shell and Hydron, Municipal Archives Amsterdam, Montevideo high-rise tower Rotterdam,

Spuimarkt The Hague etc.). Since 2006, his activities have focused on research (as part-time researcher at the TU Delft) and consulting activities and projects with his own architectural office GAAGA.

ANDREW BORGART (*1966) studied architectural design at the Rietveld Academy in Amsterdam from 1985 through 1988. From 1988 onward, he studied architecture and civil engineering at Delft University of Technology, completing his studies in 1997. Since then he has been working as an assistant professor for Structural Mechanics at the Department of Building Technology of the Faculty of Architecture of Delft University of Technology. He teaches several courses in structural mechanics as part of the Master's degree programs for Building Technology and Architectural Engineering, of which he is also the MSc coordinator. Borgart conducts research in the field of structural morphology of complex geometric structures, and in particular into the relationship between form and force of shell and membrane structures. He is joint chairman of working group 15 Structural Morphology of the International Association for Shell and Spatial Structures (IASS). He is also member of the editorial committee of the Journal of the IASS.

RAYMOND VAN SABBEN (*1982) received his Bachelor degree in Architecture from the TU Delft. In 2004 he commenced his studies in two Master's degree programs, Architecture and Building Technology, at the same institution. he graduation project entitled: "STADSkanTOREN, The design of a low-energy municipality office" was completed in early 2008. In 2005, he followed an internship at Benthem Crouwel Architects in Amsterdam, where he worked on

several projects; the design of a new interior for the reception area of the Anne Frank House in Amsterdam amongst others. After his internship he worked as a student assistant for the Chair of Design of Constructions at the Faculty of Architecture, TU Delft. Initially he worked on reorganizing the education of the Bachelor program. Since October 2006 he has been involved in various research projects of the Chair, particularly on deflateables.

P.M.J. VAN SWIETEN (*1947) graduated at the Faculty Industrial Design of the Technical University of Delft in 1978. In 1992 he was appointed assistant professor for Product Development of Building Components. He coordinates the first semester courses Component and System Design. P.M.J. van Swieten has run his own architectural practice in Leiden since 1984.

REFERENCES

M. Ashby, H. Shercliff, D. Cebon: *Materials: Engineering, science, processing and design*, Butterworth Heinemann, Oxford 2007

W. Blondeel, *Deflatable Bridge - Vacuum Construction*: MSc Thesis TU Delft, 2007

B. Burkhardt, E. Schaur, K. Bach: *IL 9 Pneus in Natur und Technik*: Institut für leichte Flächentragwerke (IL), Stuttgart 1977

B. Burkhardt, C. Thywissen, *IL 19 Growing and dividing pneus*: Institut für leichte Flächentragwerke (IL), Stuttgart 1979

R.E. Collins and T.M. Simko: *Current status of the science and technology of vacuum glazing*, School of Physics, University of Sydney

J. Gilbert, M. Patton, C. Mullen, S. Black, *Vacuumatics*: 4th year research project Queens University Department of Architecture and Planning, Belfast 1970

T. Herzog: *Pneumatic Structures*, Crosby Lockwood Staples, London 1977

F. Huijben, F. v. Herwijnen, G. Lindner: *Vacuumatic prestressed flexible architectural structures*, Structural Membranes Conference, Barcelona 2007

U. Knaack, T. Klein, M. Bilow, T. Auer: *Principles of Construction – Façades*, Birkhäuser Verlag, Berlin 2007

W. Nerdinger: *Frei Otto – leicht bauen, natürlich Gestalten*, Birkhäuser Verlag, Basle 2005

L. Nijs: *Meerlagen model*, Department Building Physics, Delft University of Technology 2001

F. Otto et al.: *Natürliche Konstruktionen*, Stuttgart 1982

A. Ritter: *Smart Materials in Architektur, Innenarchitektur und Design*, Birkhäuser Verlag, Basle 2006

C. Schittich (Hrsg.): *Gebäudehüllen – Konzepte, Schichten, Material*, Birkhäuser Verlag, Basle and Edition Detail, Munich 2001

J. Schlaich, R. Bergermann: *Leicht weit, Light structures*, Prestel Verlag, Munich 2003

W. Sobek: *Vacuumatics – Deflated forms of construction*, Detail Vol.10, pp.1148-1160, 2007

M. v. d. Voorden, L. Nijs, H. Spoorenberg: *Simulation of sound insulation properties of vacuum Zappi Façade Panels*, Seventh International IBPSA Conference, Rio de Janeiro 2001

CREDITS

IMAGINE

Series on technology and material development, Chair of Design of Constructions at Delft University of Technology. Imagine provides architects and designers with ideas and new possibilities for materials, constructions and façades by employing alternative or new technologies. It covers topics geared toward technical developments, environmental needs and aesthetic possibilities.

SERIES EDITORS
Ulrich Knaack, Tillman Klein, Marcel Bilow

PEER REVIEW
Prof. Dr Alan Brookes, Goring on Thames
Prof. Dr Gerhard Hausladen, Munich University of Technology
Prof. Kees Kaan, Delft University of Technology

DEFLATEABLES

AUTHORS
Marcel Bilow, Tillman Klein, Ulrich Knaack

WITH CONTRIBUTIONS BY
Arie Bergsma, Andrew Borgart, Raymond van Sabben, Peter van Swieten

TEXT EDITING
Usch Engelmann, Linda Hildebrand George Hall

DESIGN
Minke Themans

PRINTED BY
Die Keure, Brugge

ILLUSTRATION CREDITS
All illustrations by the authors and people who contributed to this book, except for page 15, fig. 2: reprinted by courtesy of the Otto-von-Guericke Foundation, Magdeburg, Archive; page 18, fig. 6: Imagery by Martin Tenpirrik

©2008 010 Publishers, Rotterdam
www.010publishers.nl

ISBN 978 90 6450 657 4

ALSO PUBLISHED
Imagine 01
Façades
ISBN 978 90 6450 656 7

TO BE PUBLISHED
Imagine 03
Performance Driven Envelopes

Imagine 04
Rapids